W9-DEN-103

An Introduction to Conservation of Orbital Symmetry

A programmed text

An Introduction to Conservation of Orbital Symmetry

A Programmed Text

A. J. Bellamy
Chemistry Department, University of Edinburgh

Longman

Longman Group Limited
London

Associated companies, branches and representatives
throughout the world

First published 1974

ISBN 0 582 44089 0

Set in IBM Press Roman 10 on 11 pt
and printed in Great Britain
by William Clowes & Sons Limited,
London, Colchester and Beccles

Preface

Modern organic chemistry is becoming more and more dependent upon molecular orbital theory in its search for an understanding of the behaviour of organic molecules. Probably the most fruitful interaction of these two disciplines has been the development of the concept of conservation of orbital symmetry in chemical reactions. Since its introduction by Woodward and Hoffmann in 1965, this idea has catalysed an immense amount of scientific effort, both theoretical and experimental, and it has been fascinating to follow these developments.

Unfortunately, many undergraduate and postgraduate students of organic chemistry, and also practising organic chemists, find some of the ideas involved in this approach difficult to grasp. It is therefore my sincere hope that this book, which develops the concept of conservation of orbital symmetry in the form of a programmed learning text, will help to overcome this problem, and thus extend the appreciation and application of this stimulating idea.

The Introduction sets out the basic premise and outlines the methodology which is adopted in its application. The programme itself begins with an introduction to the idea that molecular orbitals are 'built' from atomic orbitals (Chapter 1), and establishes a basis for the ideas developed in the subsequent chapters. (To those who think they already have a good understanding of this area, you may pass directly to Chapter 2, but be prepared to turn back to Chapter 1 if you are struggling!) Chapters 2, 3, and 4 introduce and develop the concept of conservation of orbital symmetry in electrocyclic reactions, in cyclo-addition reactions, and in sigmatropic migration reactions. Chapter 5 examines in more detail the conclusions reached in Chapters 2 and 3. The approach throughout is of necessity a theoretical one, but there are many questions in which the theoretical conclusions are applied to experimental results taken from the chemical literature. There is a short index at the end to enable useful tables, predictions, etc., to be quickly located.

Each section is prefaced by a *Statement* in which information and ideas are given, and this is followed by a series of *Questions* and *Answers* in which the contents of the previous Statement are examined and developed. For the reader to gain most benefit from this book, the Answers should be covered over, and should only be uncovered when the reader has answered the Question. If the response given by the reader agrees with the Answer he may pass on to the next Question. If the two do not agree, the reader should: 1) re-read the Question, 2) study the preceding Questions/Answers and/or Statement, until he understands the reasoning behind the Answer given. He may then pass on to the next Question. Periodically, the results from a series of Questions are collected together as a *Conclusion*.

I would like to thank Professor J. I. G. Cadogan for his interest and helpful advice which lead to this book being written; Dr. D. Leaver, who made some useful suggestions and also read the manuscript; Mrs. C. Ranken for her excellent typing; my colleagues and students for their discussion and searching questions, and the students who have put the programme to the test. Finally, I wish to thank my wife for her interest, patience and help.

A. J. B.

Contents

Introduction

R. B. Woodward and R. Hoffmann in 1965 (*J. Am. chem. Soc.*, 1965, **87**, 395, 2046, 2511) realised that an important factor which contributes to the activation energy of a chemical reaction is *conservation of orbital symmetry.*• They predicted that reactions which occur with conservation of orbital symmetry would have activation energies much lower than related reactions which occur without conservation of orbital symmetry. The former type of reaction, because orbital symmetry is conserved, could proceed in a concerted manner, where any bond cleavage and bond formation would take place simultaneously and continuously, resulting in strict control of the stereochemistry of the reaction. By contrast, the latter type of reaction, because orbital symmetry is not for some reason conserved, could not occur in a concerted manner. In such reactions, any bond cleavage and bond formation would have to take place in a discontinuous manner, and would probably involve an intermediate species, e.g. a diradical. Such reactions are usually not stereospecific.

These predictions, which have become known as the *Woodward-Hoffmann Rules,* and which have been amply substantiated by experiment, enable reactions to be classified as *symmetry allowed* (occurring with conservation of orbital symmetry), or *symmetry forbidden* (occurring without conservation of orbital symmetry). Some of the thermal and photochemical reactions to which this type of analysis has been successfully applied include electrocyclic reactions (see Chapter 2), cyclo-addition reactions (see Chapter 3), and sigmatropic migration reactions (see Chapter 4).

In order to deduce whether a reaction can occur with conservation of orbital symmetry, the reactant(s) and the product(s), and also all intervening species, must have a suitable element of symmetry on which to base the analysis. This may be a plane of symmetry or a two-fold axis of symmetry, and this must pass through one or more of the bonds which are broken or formed in the course of the reaction. Examination of the symmetry of the molecular orbitals which are involved in the reaction (other molecular orbitals are ignored), with respect to the symmetry element present in the reaction, enables correlations between the orbitals of the reactant(s) and product(s), based on their symmetry classifications, to be made. This leads to a prediction of the fate of the bonding electrons, and an estimate (in qualitative terms) of that part of the activation energy of the reaction which is dictated by conservation of orbital symmetry.

The systems which have been used in this book for the theoretical analysis of the subject are unsubstituted, all-carbon systems. If the conclusions arrived at were only applicable to these systems, then the concept of conservation of orbital symmetry would have very limited power, and its stereochemical predictions would be difficult to verify. Fortunately, however, the conclusions also apply to substituted systems provided the substituents are not too demanding electronically. For example, the observed stereochemical course in the electrocyclic reaction:

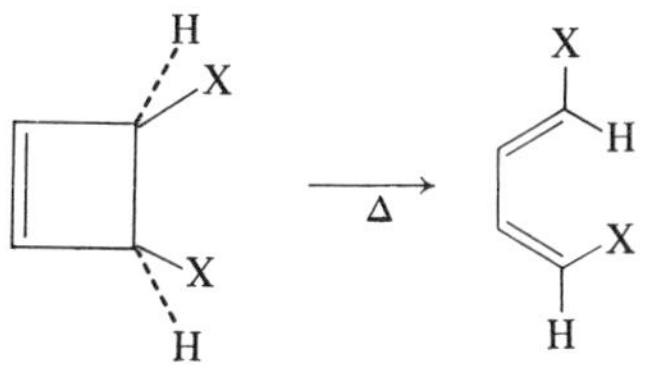

X = Me, Cl, CO_2Me

• Other important contributions to the activation energy of a chemical reaction come from changes in bond lengths, bond angles, bond strengths, non-bonded interactions, etc.

is the same as that predicted for the conversion of cyclobutene into buta-1,3-diene. Thus, although substituents obviously change the *overall* symmetry of the system, the *local* symmetry of the relevant molecular orbitals remains closely similar to that in the parent system.

One word of caution! The prediction that a reaction can occur with conservation of orbital symmetry does not necessarily mean that a concerted pathway will be followed. There may, for example, be steric restraints present in the reactant(s) or product(s) which prevent the reaction from following the stereochemical course predicted for the concerted reaction, in which case a non-concerted pathway may be preferred.

1 Atomic and molecular orbitals

Introduction

In this section of the programmed text we shall be concerned with atomic orbitals of the s, p, and s and p hybridised type, and with how these combine together to form bonds, or molecular orbitals.

S1.1 s-Type atomic orbitals are characterised by an electron distribution which is spherically symmetrical, their wave function, ϕ, has a single phase over all regions of space. p-Type atomic orbitals on the other hand have an electron distribution which is symmetrical about one of the three axes, x, y or z, and have a *nodal* plane, viz. the yz plane for a p_x-orbital, the xz plane for a p_y-orbital, and the xy plane for a p_z-orbital, at which there is a zero probability of finding an electron. s and p Hybridised orbitals also have a nodal surface. The phase of the wave function, normally represented by + and – in the orbital diagrams, becomes inverted at a nodal plane or surface.

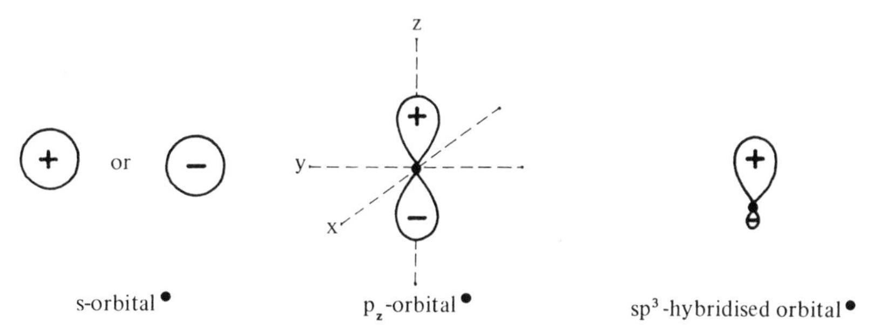

(*N.B.* 'orbitals' are not the paths in which the electrons move, but are envelopes within which there is a high (say 99%) probability of finding an electron.)

In simple terms, σ- and π-bonds are formed by the overlap of atomic orbitals, e.g.

σ-bond formation:

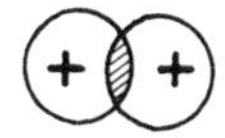

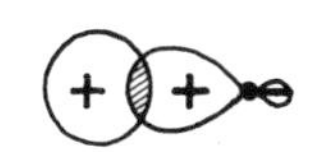

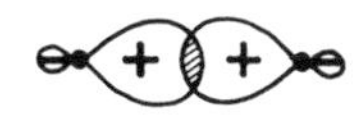

π-bond formation:

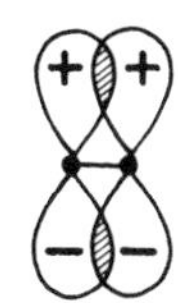

The wave functions of the orbitals resulting from the overlap of two (or more) atomic orbitals can be described by a *Linear Combination of Atomic Orbitals* (L.C.A.O.). Thus, the σ- or π-bond formed by the overlap of two atomic orbitals may be expressed as:

$$\psi = c_1\phi_1 + c_2\phi_2$$

where ψ is the wave function of the bond, ϕ_1 and ϕ_2 are the wave functions of the two atomic orbitals, and c_1 and c_2 are coefficients which give a measure of the contributions made by ϕ_1 and ϕ_2 respectively to the new wave function. The interaction of two atomic orbitals must give rise to the same number of molecular orbitals, one of which has an energy lower than either ϕ_1 or ϕ_2, and the other an energy greater than either ϕ_1 or ϕ_2. The molecular orbital of lowest energy is described by the wave function in which c_1 and c_2 are both positive (or both negative), and is a *bonding* combination of ϕ_1 and ϕ_2, while the molecular orbital of highest energy is described by the wave function in which c_1 and c_2 are of opposite sign. The latter is an *antibonding* combination of ϕ_1 and ϕ_2.

• An equivalent picture of all these atomic orbitals is obtained if the phases are reversed.

ψ antibonding

ϕ_1

ϕ_2

E

ψ bonding

$$\psi_{\text{bonding}} = (\pm c_1)\phi_1 + (\pm c_2)\phi_2 \qquad (\sigma \text{ or } \pi)$$

$$\psi_{\text{antibonding}} = (\pm c_1)\phi_1 + (\mp c_2)\phi_2 \qquad (\sigma^* \text{ or } \pi^*)$$

$$E_{\psi\ \text{bonding}} < E_{\psi\ \text{antibonding}}$$

Simple pictorial representations of the 'molecular orbitals' are shown below.

For two s-atomic orbitals:

σ-bonding

σ-antibonding (σ*)

For an s- and a p-atomic orbital:

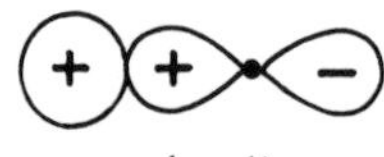

σ-bonding

σ-antibonding (σ*)

For two p-atomic orbitals;

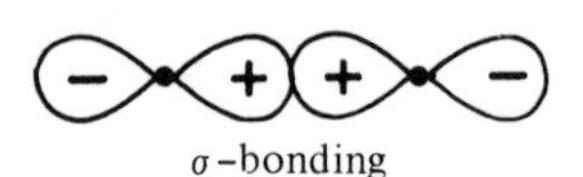

σ-bonding

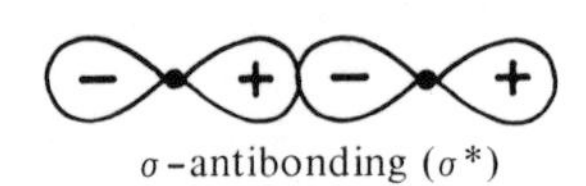

σ-antibonding (σ*)

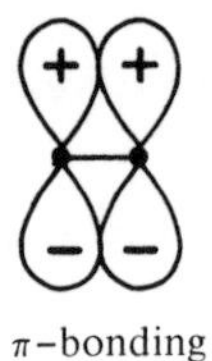

π-bonding

π-antibonding (π*)

Notice that a bonding interaction involves an 'in-phase' overlap of the atomic orbitals, i.e. a region in which the phases of ϕ_1 and ϕ_2 are either both positive or both negative; and that an antibonding interaction involves an 'out-of-phase' overlap of the atomic orbitals, i.e. a region in which the phases of ϕ_1 and ϕ_2 are different. Notice also that there is a nodal surface between the two atoms in an antibonding orbital.

Generally, a σ-type of interaction of two atomic orbitals produces a bonding molecular orbital of lower energy, and an antibonding molecular orbital of higher energy, than a π-type of interaction for the same internuclear distance, e.g. for two p-orbitals:

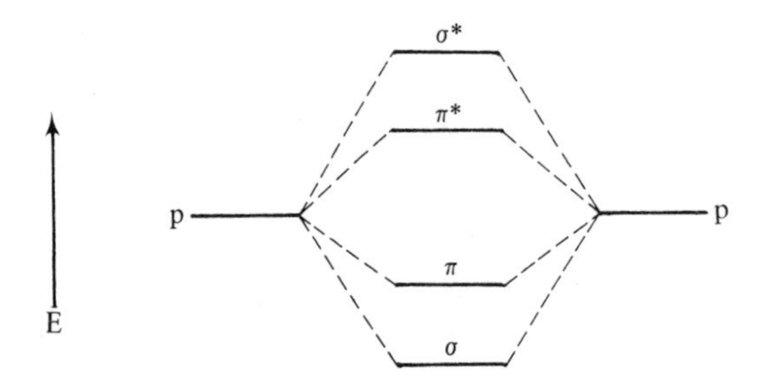

Which molecular orbitals contain electrons depends upon the total number of electrons to be accommodated, and upon whether the system is in its electronic ground-state or an electronic excited-state. In the electronic ground-state, the electrons occupy the orbitals of lowest energy, each orbital accommodating a maximum of two electrons.

Q1.1 Which will be the occupied molecular orbitals in the electronic ground-states of the following species: (a) H_2^+, (b) H_2, (c) H_2^-? Indicate how many electrons will be in each orbital.

A1.1 (a) σ^1 (b) σ^2 (c) $\sigma^2\sigma^{*1}$

The number of electrons in an orbital is usually represented by a superscript after the symbol for the orbital. Thus in (c), the σ-orbital contains two electrons and the σ*-orbital contains one electron

Q1.2 For the two types of carbon-carbon bond in ethylene (σ and π), arrange the molecular orbitals in order of increasing energy.

A1.2

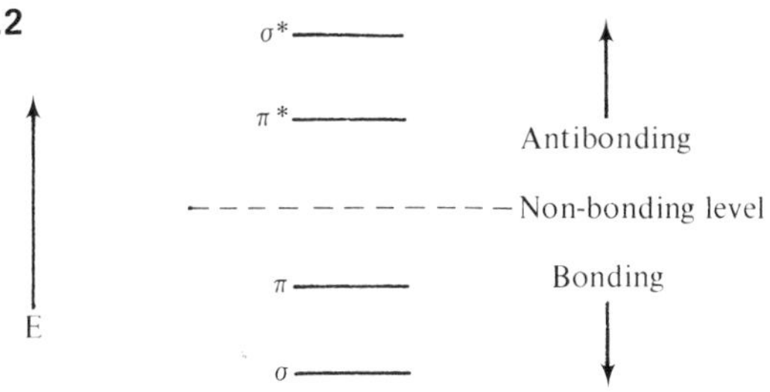

Q1.3 Which molecular orbitals of the carbon–carbon bond will be occupied in the electronic ground-state of ethylene?

A1.3 $\sigma^2\pi^2$

Q1.4 For the electronic ground-state of $[CH_2{=}CH_2]^+$, which orbitals will be occupied? Indicate the number of electrons in each orbital.

A1.4 $\sigma^2\pi^1$

Q1.5 For the electronic ground-state of $[CH_2{=}CH_2]^{\overline{\cdot}}$, which orbitals will be occupied? Indicate the number of electrons in each orbital.

A1.5 $\sigma^2\pi^2\pi^{*1}$

S1.2 The absorption of energy in the ultraviolet region of the electromagnetic radiation spectrum causes an electron in an occupied molecular orbital to be promoted into an unoccupied molecular orbital of higher energy. This produces an electronic excited-state of the molecule.

Q1.6 Which electronic excitation of ethylene requires least energy?

A1.6 The occupied orbitals are: $\sigma^2\pi^2$
The unoccupied orbitals are: π^* and σ^*
The smallest energy transition is $\pi \rightarrow \pi^*$

Q1.7 Which electronic configuration of ethylene is produced by a $\pi \rightarrow \pi^*$ transition?

A1.7 $\sigma^2\pi^2 \rightarrow \sigma^2\pi^1\pi^{*1}$

S1.3 We have examined the π-type of interaction between two p-orbitals on adjacent (carbon) atoms. Before proceeding further, we need to examine the molecular orbitals which result when there are p-orbitals on more than two consecutive carbon atoms:

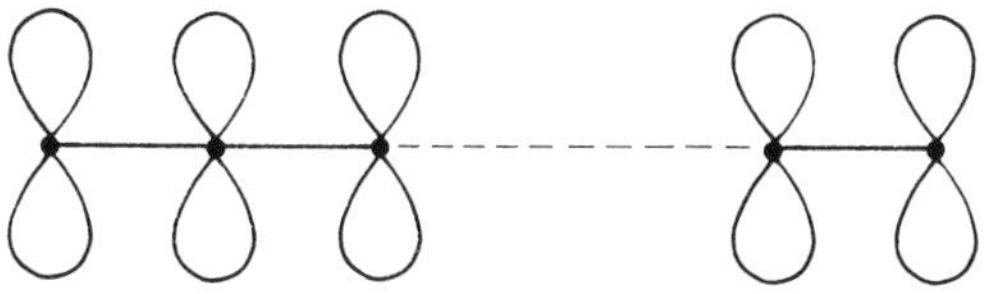

The wave functions of the molecular orbitals which result from the delocalisation of the electrons in these p-orbitals may again be described by a Linear Combination of Atomic Orbitals (L.C.A.O.):

$$\psi = c_1\phi_1 + c_2\phi_2 + c_3\phi_3 + \cdots + c_n\phi_n$$

and if there are n p-orbitals, then n molecular orbitals will result:

$$\psi_1 = c_{11}\phi_1 + c_{21}\phi_2 + c_{31}\phi_3 + \cdots + c_{n1}\phi_n$$

$$\psi_2 = c_{12}\phi_1 + c_{22}\phi_2 + c_{32}\phi_3 + \cdots + c_{n2}\phi_n$$

$$\psi_n = c_{1n}\phi_1 + c_{2n}\phi_2 + c_{3n}\phi_3 + \cdots + c_{nn}\phi_n$$

The contribution made by each atomic orbital in a particular molecular orbital is given by the sign and magnitude of the coefficient, c, of the atomic orbital in the wave function for the molecular orbital.

The values of the coefficients are obtained mathematically by satisfying certain conditions, viz. the energy of the resulting molecular orbitals must be a minimum, and the probability of finding an electron with the wave function ψ in all space must be unity. The details of the method may be found in most textbooks on this subject (see References on p. 10).

This approximate treatment, introduced by E. Hückel in 1931, gives reasonable values for the energy of an electron in each molecular orbital, and the coefficients of the atomic orbitals in the wave functions. The energy values are not expressed in absolute units, but in terms of two constants, α and β.

$$E = \alpha + m\beta$$

α is the energy of an electron localised in a 2p-atomic orbital of carbon (known as the Coulomb integral), and β is the energy of a π-type of interaction between two adjacent 2p-orbitals (known as the Resonance integral). Both α and β are negative energy quantities (relative to an electron at infinity), therefore $E = \alpha + \beta$ is of lower energy than $E = \alpha - \beta$.

For linear polyenes with n atoms:

$$m \text{ (for the } j\text{th molecular orbital)} = 2\cos\frac{j\pi}{(n+1)}$$

$$(j = 1, 2, 3, \ldots n)$$

$$E_j = \alpha + \left(2\cos\frac{j\pi}{n+1}\right)\beta \qquad \text{Equation 1}^{\bullet}$$

The coefficients for the contributions made by the atomic orbitals in the jth molecular orbital can be calculated from the expression:

$$c_{rj} = \left(\frac{2}{n+1}\right)^{\frac{1}{2}} . \sin\frac{rj\pi}{n+1} \qquad \text{Equation 2}^{\bullet}$$

where c_{rj} is the coefficient of the rth atomic orbital in the jth molecular orbital.

• These general equations were derived by C. A. Coulson, and are given in: C. A. Coulson and H. C. Longuet-Higgins, *Proc. Roy. Soc.*, 1947, **A192**, 16.

Q1.8 Using equations 1 and 2, calculate the energies of the two π-molecular orbitals of ethylene, and their wave functions.

A1.8
$$E_1 = \alpha + \left(2\cos\frac{\pi}{3}\right)\beta = \alpha + \beta \quad \text{(bonding)}$$
$$E_2 = \alpha + \left(2\cos\frac{2\pi}{3}\right)\beta = \alpha - \beta \quad \text{(antibonding)}$$
$$c_{11} = \left(\frac{2}{3}\right)^{\frac{1}{2}} . \sin\frac{\pi}{3} = \frac{1}{2^{\frac{1}{2}}} = 0.71$$
$$c_{21} = \left(\frac{2}{3}\right)^{\frac{1}{2}} . \sin\frac{2\pi}{3} = \frac{1}{2^{\frac{1}{2}}} = 0.71$$
$$\psi_1 = 0.71\,(\phi_1 + \phi_2) \quad \text{(bonding)}$$
$$c_{12} = \left(\frac{2}{3}\right)^{\frac{1}{2}} . \sin\frac{2\pi}{3} = \frac{1}{2^{\frac{1}{2}}} = 0.71$$
$$c_{22} = \left(\frac{2}{3}\right)^{\frac{1}{2}} . \sin\frac{4\pi}{3} = -\frac{1}{2^{\frac{1}{2}}} = -0.71$$
$$\psi_2 = 0.71\,(\phi_1 - \phi_2) \quad \text{(antibonding)}$$

Q1.9 Calculate the energies of the three π-molecular orbitals of the prop-2-enyl (allyl) system,

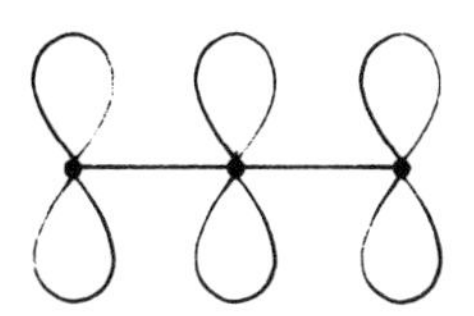

and their wave functions.

A1.9
$$E_1 = \alpha + \left(2\cos\frac{\pi}{4}\right)\beta = \alpha + 1.414\,\beta$$
$$E_2 = \alpha + \left(2\cos\frac{2\pi}{4}\right)\beta = \alpha$$
$$E_3 = \alpha + \left(2\cos\frac{3\pi}{4}\right)\beta = \alpha - 1.414\,\beta$$
$$c_{11} = \left(\frac{2}{4}\right)^{\frac{1}{2}} . \sin\frac{\pi}{4} = \frac{1}{2} = 0.50$$
$$c_{21} = \left(\frac{2}{4}\right)^{\frac{1}{2}} . \sin\frac{2\pi}{4} = \frac{1}{2^{\frac{1}{2}}} = 0.71$$
$$c_{31} = \left(\frac{2}{4}\right)^{\frac{1}{2}} . \sin\frac{3\pi}{4} = \frac{1}{2} = 0.50$$
$$\psi_1 = 0.50\phi_1 + 0.71\phi_2 + 0.50\phi_3$$
$$c_{12} = \left(\frac{2}{4}\right)^{\frac{1}{2}} . \sin\frac{2\pi}{4} = \frac{1}{2^{\frac{1}{2}}} = 0.71$$
$$c_{22} = \left(\frac{2}{4}\right)^{\frac{1}{2}} . \sin\frac{4\pi}{4} = 0$$
$$c_{32} = \left(\frac{2}{4}\right)^{\frac{1}{2}} . \sin\frac{6\pi}{4} = -\frac{1}{2^{\frac{1}{2}}} = -0.71$$
$$\psi_2 = 0.71\,(\phi_1 - \phi_3)$$
$$c_{13} = \left(\frac{2}{4}\right)^{\frac{1}{2}} . \sin\frac{3\pi}{4} = \frac{1}{2} = 0.50$$
$$c_{23} = \left(\frac{2}{4}\right)^{\frac{1}{2}} . \sin\frac{6\pi}{4} = -\frac{1}{2^{\frac{1}{2}}} = -0.71$$
$$c_{33} = \left(\frac{2}{4}\right)^{\frac{1}{2}} . \sin\frac{9\pi}{4} = \frac{1}{2} = 0.50$$
$$\psi_3 = 0.50\phi_1 - 0.71\phi_2 + 0.50\phi_3$$

Q1.10 Calculate the wave functions and their energies for the four π-molecular orbitals of buta-1,3-diene.

A1.10 $\psi_1 = 0.37\phi_1 + 0.60\phi_2 + 0.60\phi_3 + 0.37\phi_4$

$\psi_2 = 0.60\phi_1 + 0.37\phi_2 - 0.37\phi_3 - 0.60\phi_4$

$\psi_3 = 0.60\phi_1 - 0.37\phi_2 - 0.37\phi_3 + 0.60\phi_4$

$\psi_4 = 0.37\phi_1 - 0.60\phi_2 + 0.60\phi_3 - 0.37\phi_4$

$E_1 = \alpha + 1.618\beta$

$E_2 = \alpha + 0.618\beta$

$E_3 = \alpha - 0.618\beta$

$E_4 = \alpha - 1.618\beta$

S1.4 If the energy of a molecular orbital is less than α, then that particular orbital is a bonding molecular orbital, e.g. ψ_1 ($E_1 = \alpha + 1.618\beta$) and ψ_2 ($E_2 = \alpha + 0.618\beta$) of buta-1,3-diene. Conversely, if the energy is greater than α, then the orbital is an antibonding molecular orbital, e.g. ψ_3 ($E_3 = \alpha - 0.618\beta$) and ψ_4 ($E_4 = \alpha - 1.618\beta$) of buta-1,3-diene. (Remember that α and β are negative energy quantities relative to an electron at infinity.)

Q1.11 For the molecular orbitals of the prop-2-enyl system, say which orbital is a bonding molecular orbital, and which orbital is an antibonding molecular orbital.

A1.11 ψ_1 is a bonding orbital ($E_1 = \alpha + 1.414\beta$; $E_1 < \alpha$). ψ_3 is an antibonding orbital ($E_3 = \alpha - 1.414\beta$; $E_3 > \alpha$).

S1.5 The coefficients of the atomic orbitals in the wave functions of the molecular orbitals provide the following information:

(a) the magnitude of the contribution which a particular atomic orbital makes in a given molecular orbital,

(b) the existence of a bonding, antibonding, or non-bonding situation between two adjacent atoms,

(c) the presence and position of nodes.

For example, in ψ_2 of buta-1,3-diene:

(a) ϕ_1 and ϕ_4 make a larger contribution than ϕ_2 and ϕ_3 ($|c_1|$ and $|c_4| > |c_2|$ and $|c_3|$),

(b) a bonding situation exists between $C_{(1)}$ and $C_{(2)}$, and between $C_{(3)}$ and $C_{(4)}$ (c_1 and c_2 both positive; c_3 and c_4 both negative), and an antibonding situation exists between $C_{(2)}$ and $C_{(3)}$ (c_2 and c_3 have opposite sign),

(c) a node (change of phase in the wave function) is present between $C_{(2)}$ and $C_{(3)}$.

ψ_2 of the prop-2-enyl system is a non-bonding orbital. ϕ_2 makes no contribution to the wave function ($c_2 = 0$; there is a node at $C_{(2)}$), and a non-bonding situation exists between $C_{(1)}$ and $C_{(2)}$, and between $C_{(2)}$ and $C_{(3)}$. Electrons in this orbital have the same energy as they would in a localised 2p-atomic orbital, viz. α.

Q1.12 By examining the coefficients of the atomic orbitals in ψ_3 of buta-1,3-diene, deduce where (a) a bonding, and (b) an antibonding situation exists between adjacent atoms.

A1.12 $\psi_3 = 0.60\phi_1 - 0.37\phi_2 - 0.37\phi_3 + 0.60\phi_4$.

There is bonding between $C_{(2)}$ and $C_{(3)}$ (c_2 and c_3 both negative).

There is antibonding between $C_{(1)}$ and $C_{(2)}$, and between $C_{(3)}$ and $C_{(4)}$ (c_1 and c_2, and c_3 and c_4, have opposite sign).

Q1.13 How many nodes have each of the molecular orbitals of buta-1,3-diene? Indicate their position.

A1.13 ψ_1: no node.
ψ_2: one node between $C_{(2)}$ and $C_{(3)}$.
ψ_3: two nodes, one between $C_{(1)}$ and $C_{(2)}$, and one between $C_{(3)}$ and $C_{(4)}$.
ψ_4: three nodes, one between each pair of adjacent atoms.

Q1.14 How many nodes have each of the molecular orbitals of the prop-2-enyl system? Indicate their position.

A1.14 ψ_1: no node.
ψ_2: one node at $C_{(2)}$.
ψ_3: two nodes, one between each pair of adjacent atoms.

Q1.15 Does the energy of a molecular orbital decrease or increase with an increase in the number of nodes?

Q1.15 The energy increases with the number of nodes.

Q1.16 π-Molecular orbitals may be represented pictorially in the following way: for ψ_2 of buta-1,3-diene,

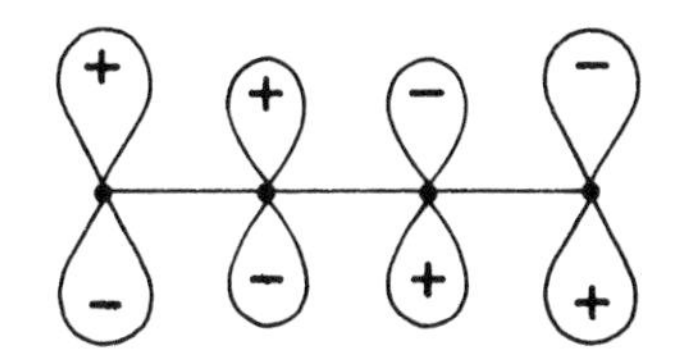

The orbitals at $C_{(1)}$ and $C_{(4)}$ are drawn larger than those at $C_{(2)}$ and $C_{(3)}$ because the coefficients, and hence the contributions, of ϕ_1 and ϕ_4 in ψ_2 are larger than those of ϕ_2 and ϕ_3.

Draw similar representations of π and π^* of ethylene, ψ_1, ψ_2, and ψ_3 of the prop-2-enyl system, and ψ_1, ψ_3, and ψ_4 of buta-1,3-diene.

A1.16 Ethylene:

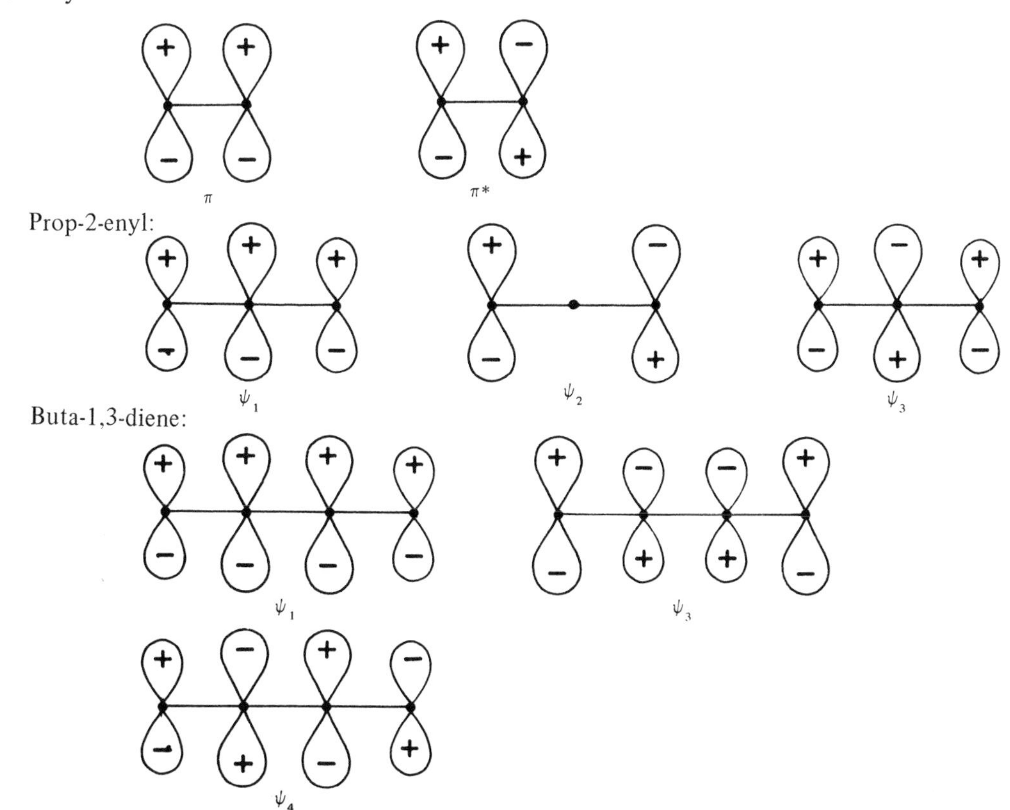

S1.6 For an analysis of conservation of orbital symmetry in chemical reactions, we are not too concerned that ϕ_2 and ϕ_3 make a smaller contribution than ϕ_1 and ϕ_4 to the wave functions ψ_2 and ψ_3 of buta-1,3-diene. What we are more interested in is the phase relation between the various atomic orbitals in each molecular orbital, and the nodal properties of the latter. Therefore, from now on, all atomic orbitals in the pictorial representation of molecular orbitals will be drawn the same size, e.g. ψ_2 of buta-1,3-diene:

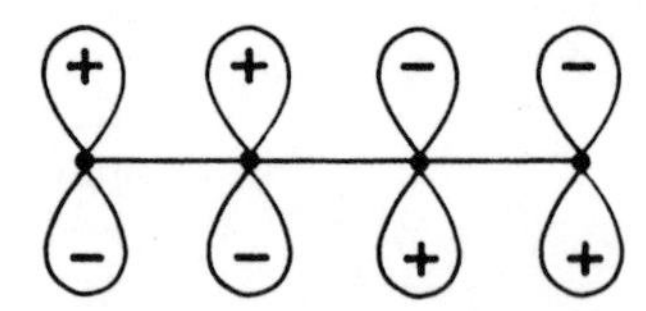

It is convenient to be able to represent molecular orbitals in a more abbreviated form, while retaining as much of the useful information contained in the original form as possible. One method which satisfies these requirements is:

for ψ_2 of buta-1,3-diene   + + – –

This gives the relative phases of the upper lobes of the atomic orbitals in ψ_2, and shows immediately where there is bonding and antibonding, and the position of a node. Similarly, ψ_2 of the prop-2-enyl system may be written as: + 0 –.

The Table below gives a list of the π-molecular orbitals of the linear polyenes up to C_8 written in this way, together with their energies, the latter increasing from the bottom.

TABLE

No. of carbon atoms in polyene	Wave function		Energy
C_2	ψ_2	+ –	$\alpha - \beta$
	ψ_1	+ +	$\alpha + \beta$
C_3	ψ_3	+ – +	$\alpha - 1.414\beta$
	ψ_2	+ 0 –	α
	ψ_1	+ + +	$\alpha + 1.414\beta$
C_4	ψ_4	+ – + –	$\alpha - 1.618\beta$
	ψ_3	+ – – +	$\alpha - 0.618\beta$
	ψ_2	+ + – –	$\alpha + 0.618\beta$
	ψ_1	+ + + +	$\alpha + 1.618\beta$
C_5	ψ_5	+ – + – +	$\alpha - 1.732\beta$
	ψ_4	+ – 0 + –	$\alpha - \beta$
	ψ_3	+ 0 – 0 +	α
	ψ_2	+ + 0 – –	$\alpha + \beta$
	ψ_1	+ + + + +	$\alpha + 1.732\beta$
C_6	ψ_6	+ – + – + –	$\alpha - 1.802\beta$
	ψ_5	+ – + + – +	$\alpha - 1.247\beta$
	ψ_4	+ – – + + –	$\alpha - 0.445\beta$
	ψ_3	+ + – – + +	$\alpha + 0.445\beta$
	ψ_2	+ + + – – –	$\alpha + 1.247\beta$
	ψ_1	+ + + + + +	$\alpha + 1.802\beta$
C_7	ψ_7	+ – + – + – +	$\alpha - 1.848\beta$
	ψ_6	+ – + 0 – + –	$\alpha - 1.414\beta$
	ψ_5	+ – – + – – +	$\alpha - 0.765\beta$
	ψ_4	+ 0 – 0 + 0 –	α
	ψ_3	+ + – – – + +	$\alpha + 0.765\beta$
	ψ_2	+ + + 0 – – –	$\alpha + 1.414\beta$
	ψ_1	+ + + + + + +	$\alpha + 1.848\beta$

TABLE (*continued*)

C_8	ψ_8	+	−	+	−	+	−	+	−	$\alpha - 1.879\beta$
	ψ_7	+	−	+	−	−	+	−	+	$\alpha - 1.532\beta$
	ψ_6	+	−	0	+	−	0	+	−	$\alpha - \beta$
	ψ_5	+	−	−	+	+	−	−	+	$\alpha - 0.347\beta$
	ψ_4	+	+	−	−	+	+	−	−	$\alpha + 0.347\beta$
	ψ_3	+	+	0	−	−	0	+	+	$\alpha + \beta$
	ψ_2	+	+	+	+	−	−	−	−	$\alpha + 1.532\beta$
	ψ_1	+	+	+	+	+	+	+	+	$\alpha + 1.879\beta$

Q1.17 From the Table, deduce between which adjacent atoms there is a bonding, non-bonding, or antibonding situation for (a) ψ_4 of hexa-1,3,5-triene, (b) ψ_2 of the penta-2,4-dienyl system, and (c) ψ_6 of the hepta-2,4,6-trienyl system.

A1.17 (a) Bonding between $C_{(2)}$ and $C_{(3)}$, and $C_{(4)}$ and $C_{(5)}$. Antibonding between $C_{(1)}$ and $C_{(2)}$, $C_{(3)}$ and $C_{(4)}$, and $C_{(5)}$ and $C_{(6)}$.

(b) Bonding between $C_{(1)}$ and $C_{(2)}$, and $C_{(4)}$ and $C_{(5)}$. Non-bonding between $C_{(2)}$ and $C_{(3)}$, and $C_{(3)}$ and $C_{(4)}$.

(c) Non-bonding between $C_{(3)}$ and $C_{(4)}$, and $C_{(4)}$ and $C_{(5)}$. Antibonding between $C_{(1)}$ and $C_{(2)}$, $C_{(2)}$ and $C_{(3)}$, $C_{(5)}$ and $C_{(6)}$, and $C_{(6)}$ and $C_{(7)}$.

Q1.18 Using the Table, draw pictorial representations of (a) ψ_3 of hexa-1,3,5-triene, and (b) ψ_4 of the penta-2,4-dienyl system.

A1.18

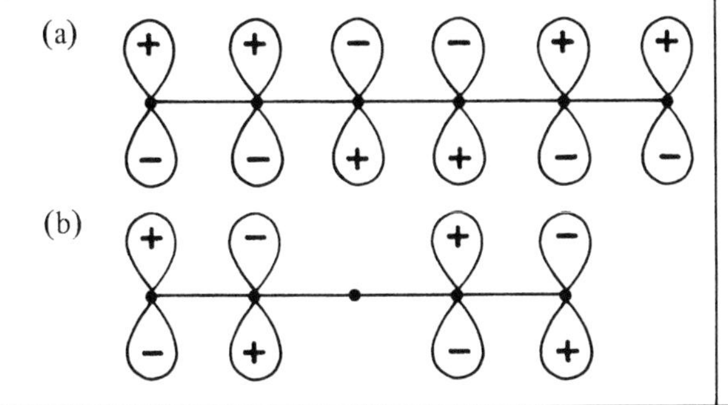

Q1.19 From the Table, deduce how many nodes the following orbitals have: (a) ψ_5 of hexa-1,3,5-triene, (b) ψ_6 of octa-1,3,5,7-tetraene.

A1.19 (a) Four (b) Five

Q1.20 What is the electronic configuration of the ground-state of the following systems: (a) buta-1,3-diene, (b) hexa-1,3,5-triene, (c) prop-2-enyl anion, (d) penta-2,4-dienyl radical, (e) hepta-2,4,6-trienyl radical?

A1.20 (a) $\psi_1^2\psi_2^2$ (b) $\psi_1^2\psi_2^2\psi_3^2$ (c) $\psi_1^2\psi_2^2$ (d) $\psi_1^2\psi_2^2\psi_3^1$ (e) $\psi_1^2\psi_2^2\psi_3^2\psi_4^1$

Q1.21 In the electronic excitation of ground-state buta-1,3-diene, which transition requires the smallest amount of energy? Which electronic configuration of buta-1,3-diene would be produced by this transition?

A1.21 $\psi_2 \rightarrow \psi_3$ requires least energy (1.236β). The other possible transitions, $\psi_2 \rightarrow \psi_4$, $\psi_1 \rightarrow \psi_3$, and $\psi_1 \rightarrow \psi_4$, require 2.236$\beta$, 2.236$\beta$, and 3.236$\beta$ units of energy respectively.

$\psi_1^2\psi_2^2 \xrightarrow{\psi_2 \rightarrow \psi_3} \psi_1^2\psi_2^1\psi_3^1$ (the 1st excited-state of buta-1,3-diene).

Q1.22 What is the electronic configuration of the 1st excited-state of the following systems: (a) prop-2-enyl anion, (b) penta-2,4-dienyl cation, (c) hexa-1,3,5-triene? What electronic transitions are involved in forming these states from the corresponding ground-states, and how much energy (in units of β) is absorbed?

A1.22 (a) $\psi_1^2\psi_2^1\psi_3^1$ ($\psi_2 \rightarrow \psi_3$; 1.414$\beta$)

(b) $\psi_1^2\psi_2^1\psi_3^1$ ($\psi_2 \rightarrow \psi_3$; β)

(c) $\psi_1^2\psi_2^2\psi_3^1\psi_4^1$ ($\psi_3 \rightarrow \psi_4$; 0.89$\beta$).

Q1.23 Which of the two systems, ethylene and buta-1,3-diene, requires the least energy to attain its 1st excited-state?

A1.23 Ethylene: $\psi_1^2 \rightarrow \psi_1^1\psi_2^1$ (2β)

Buta-1,3-diene: $\psi_1^2\psi_2^2 \rightarrow \psi_1^2\psi_2^1\psi_3^1$ (1.236β).

∴ Buta-1,3-diene requires less energy than ethylene to reach its 1st excited-state.

Q1.24 Give suggestions for the electronic configuration of the 2nd excited-state of hexa-1,3,5-triene, formed by the promotion of *one* electron from the ground-state. What electronic transitions are involved in forming these states, and how much energy (in units of β) is absorbed?

A1.24 $\psi_1^2\psi_2^1\psi_3^2\psi_4^1$ ($\psi_2 \rightarrow \psi_4$, 1.692$\beta$)

$\psi_1^2\psi_2^2\psi_3^1\psi_5^1$ ($\psi_3 \rightarrow \psi_5$; 1.692$\beta$).

REFERENCES

Roberts, J. D., *Notes on Molecular Orbital Calculations,* Benjamin, New York, 1962.
Streitwieser, A, Jr., *Molecular Orbital Theory for Organic Chemists,* Wiley, New York, 1961.
Murrell, J. N., Kettle, S. F. A. and Tedder, J. M., *Valence Theory*, Wiley, London, 1965.
Coulson, C. A., *Valence*, Oxford University Press, 1961.
Wiberg, K. B. *Physical Organic Chemistry,* Wiley, New York, 1964.
Salem, L., *The Molecular Orbital Theory of Conjugated Systems,* Benjamin, New York, 1966.
Coulson, C. A. and Streitwieser, A. Jr., *Dictionary of π-Electron Calculations,* Pergamon, 1965.

2 Electrocyclic reactions

Introduction

An electrocyclic reaction involves the formation of a σ-bond between the ends of a linear π-system, or the reverse:

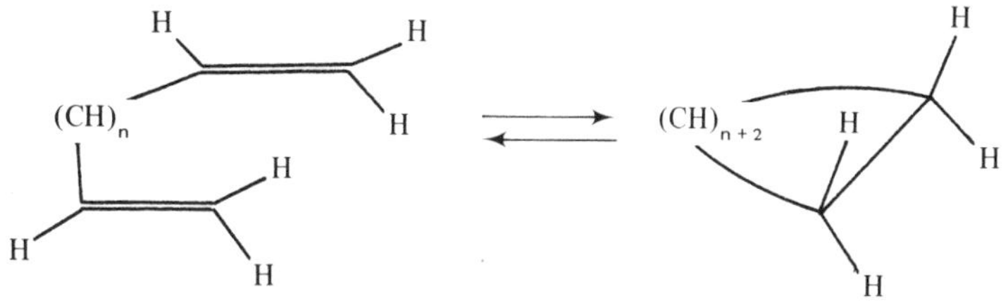

$$(CH)_n \text{ (diene termini, H substituents)} \rightleftharpoons (CH)_{n+2} \text{ (ring)}$$

For example, the thermal conversion of *trans, cis, trans*-octa-2,4,6-triene into *cis*-5,6-dimethylcyclohexa-1,3-diene:

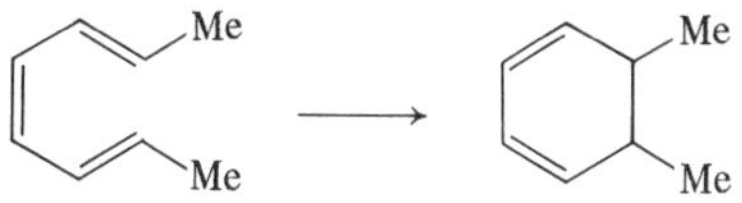

Me Me → Me Me

The analysis of this class of reactions for conservation of orbital symmetry will be dealt with at three levels of sophistication, (a) a Frontier Orbital approach, (b) using Orbital Correlation Diagrams, and (c) using State Correlation Diagrams (Chapter 5).

Frontier Orbital Approach

S2.1 This approach considers *only* the molecular orbital of highest energy which contains electrons (the Frontier Orbital), and follows the fate of the electrons in this orbital during the chemical change cf. the simple picture of the reactions of atoms where only the outer shell (valency) electrons of atoms are considered to be involved. In both contexts, these electrons are those of highest energy and therefore are most easily reorganised.

Q2.1 Name and draw the phase relations in the highest occupied molecular orbital of buta-1,3-diene in its ground-state.

A2.1 Ground-state buta-1,3-diene has 4 π-electrons, 2 in ψ_1 and 2 in ψ_2. ψ_2 is therefore the occupied molecular orbital of highest energy.

(+) (+) (−) (−) / (−) (−) (+) (+) ψ_2

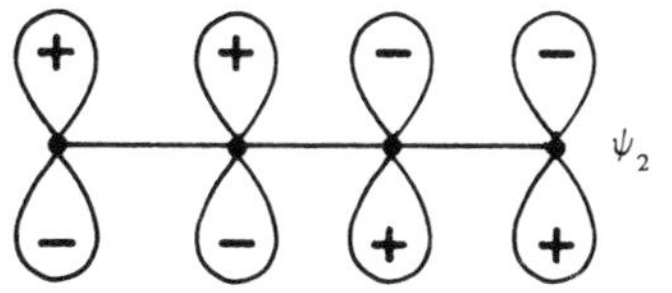

Q2.2 Name and draw the phase relations in the highest occupied molecular orbital of buta-1,3-diene in its 1st excited-state.

A2.2 Buta-1,3-diene in its 1st excited-state has 4 π-electrons, 2 in ψ_1, 1 in ψ_2, and 1 in ψ_3. ψ_3 is therefore the occupied molecular orbital of highest energy.

(+) (−) (−) (+) / (−) (+) (+) (−) ψ_3

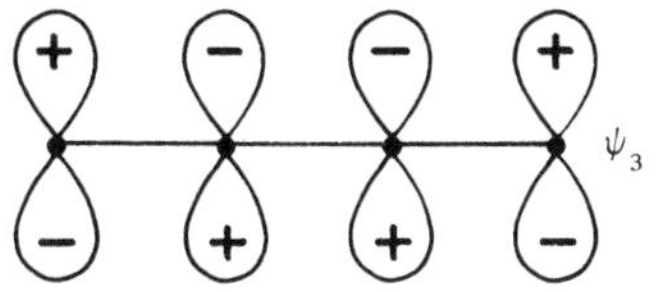

Q2.3 Give the highest occupied molecular orbital for the following systems in (i) their ground-state, and (ii) their 1st excited-state: (a) hexa-1,3,5-triene, (b) prop-2-enyl cation $[CH_2\text{⋯}CH\text{⋯}CH_2]^+$, (c) prop-2-enyl anion $[CH_2\text{⋯}CH\text{⋯}CH_2]^-$.

A2.3 (a) (i) ψ_3 (ii) ψ_4
(b) (i) ψ_1 (ii) ψ_2
(c) (i) ψ_2 (ii) ψ_3

S2.2 In the thermal isomerisation of buta-1,3-diene to cyclobutene, a σ-bond is formed between $C_{(1)}$ and $C_{(4)}$ of butadiene.

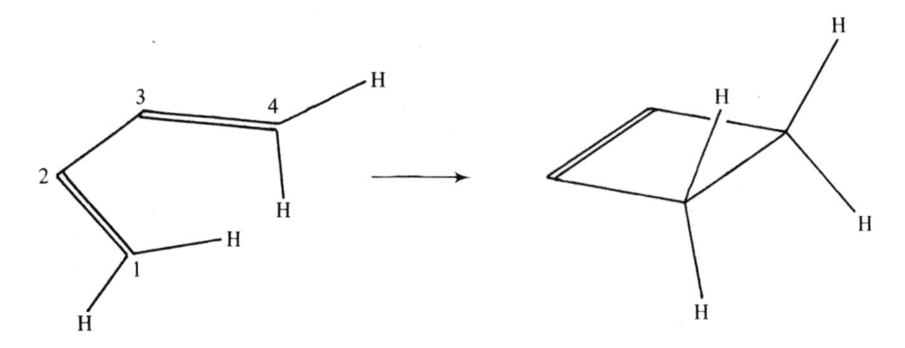

Initially, the substituents, i.e. hydrogen, at $C_{(1)}$ and $C_{(4)}$ of the diene, lie in the plane of the carbon atoms, whereas in the cyclised product the same substituents lie above and below the plane of the carbon atoms. This transformation is achieved by rotation about the $C_{(1)}C_{(2)}$ and $C_{(3)}C_{(4)}$ bonds of the diene.

The direction of rotation about these bonds may be predicted by consideration of the highest occupied molecular orbital of butadiene, ψ_2. Rotation of both bonds in the same direction, called CONROTATORY rotation, produces from ψ_2 a σ-bonding situation between $C_{(1)}$ and $C_{(4)}$ of the diene,

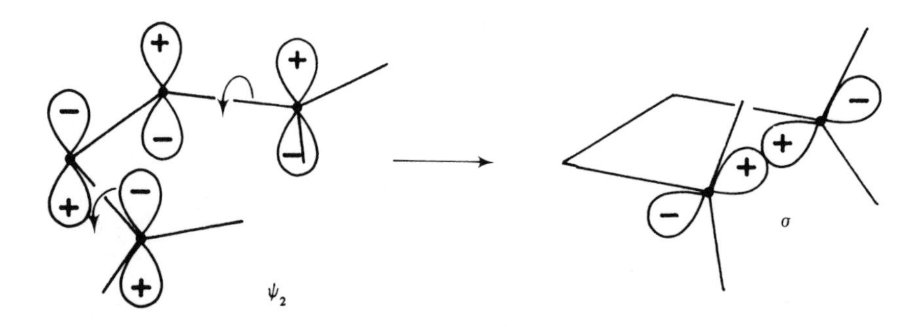

whereas rotation of the bonds in opposite directions, called DISROTATORY rotation, produces a σ-antibonding situation between $C_{(1)}$ and $C_{(4)}$.

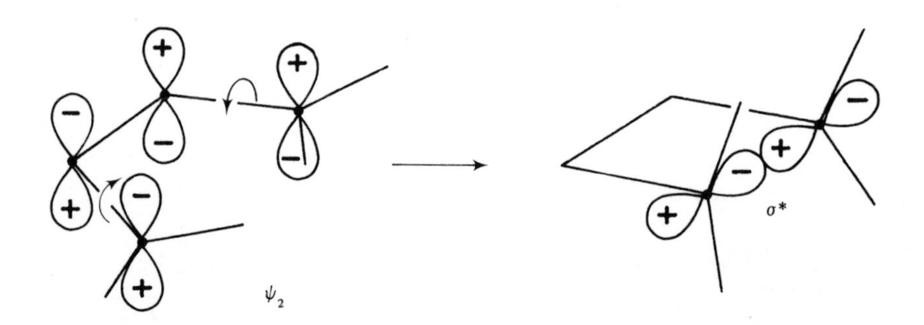

Q2.4 There are always two possible conrotatory modes of rotation, one in which both bonds rotate in an anticlockwise direction, as shown in the Statement, and one in which both bonds rotate in a clockwise direction. Show for the latter case that the same net result is obtained as for the former case, i.e. that both modes in an unsubstituted system are equivalent.

A2.4

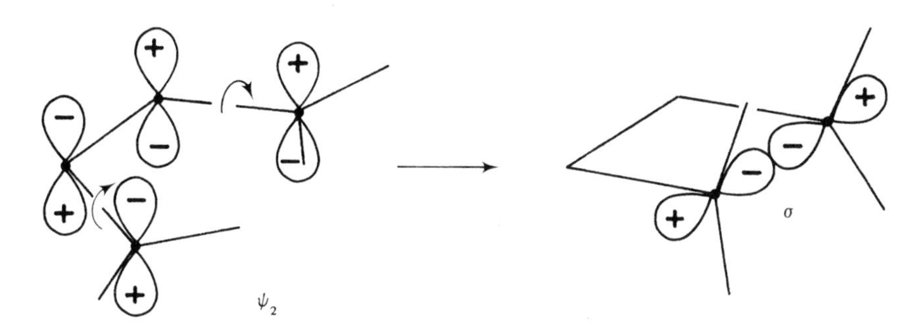

Q2.5 There are always two possible disrotatory modes of rotation. One of these is given in the Statement; the other possibility involves a reversal in the direction of rotation of both bonds. Show for the latter case that the same net result is obtained as for the former case, i.e. that both modes in an unsubstituted system are equivalent.

A2.5

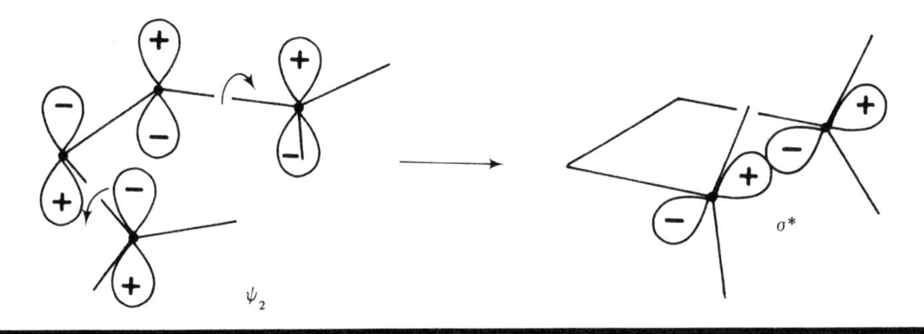

C2.1 Conrotatory rotation of ψ_2 in butadiene leads to σ in cyclobutene. Thus, two bonding electrons in butadiene are transformed into two bonding electrons in cyclobutene.

Disrotatory rotation of ψ_2 in butadiene places the same two electrons in an orbital of much higher energy. Hence, for a thermal transformation, in which ground-state electronic configurations are involved, the conrotatory process will be favoured.

Q2.6 In the photochemical isomerisation of buta-1,3-diene to cyclobutene, involving reaction from the 1st excited-state of butadiene, say which mode of rotation, conrotatory or disrotatory, in the highest occupied molecular orbital has to occur to produce a σ-bonding situation between $C_{(1)}$ and $C_{(4)}$.

A2.6 The highest occupied molecular orbital in the 1st excited-state of butadiene is ψ_3.

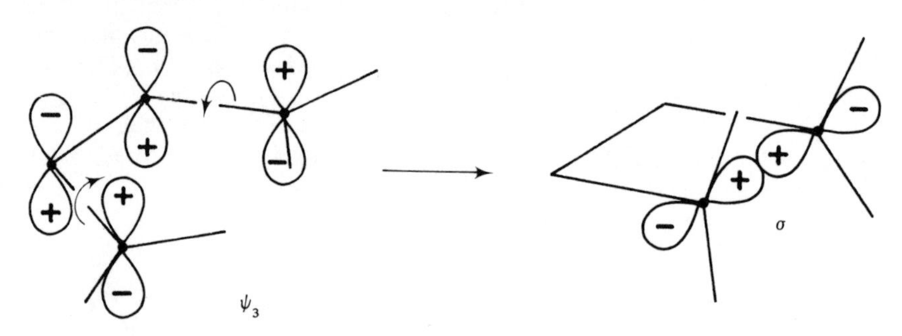

Disrotatory rotation produces a σ-bonding situation. Conrotatory rotation produces a σ-antibonding situation. Hence, the former is the preferred process.

Q2.7 to 2.12 By considering the highest occupied molecular orbital (H. O. M. O.) of each system, predict which mode of rotation is required to achieve a σ-bonding situation between the termini of the following π-systems.

Q2.7 *cis*-Hexa-1,3,5-triene in (i) its ground-state, and (ii) its 1st excited-state.

A2.7 (i) Electronic configuration: $\psi_1^2\psi_2^2\psi_3^2$
ψ_3 is the H. O. M. O.

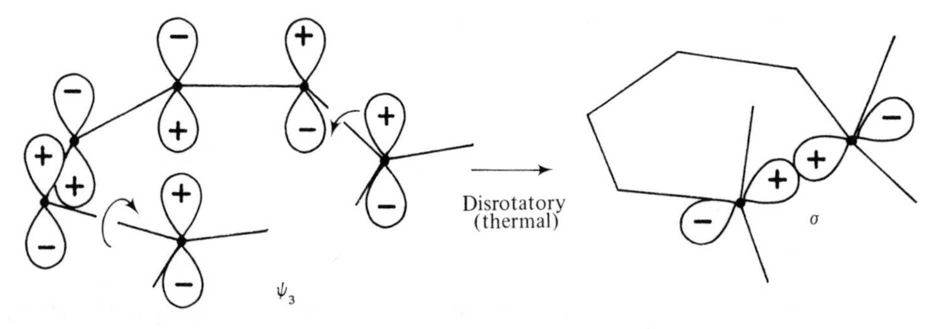

(ii) Electronic configuration: $\psi_1^2\psi_2^2\psi_3^1\psi_4^1$
ψ_4 is the H.O.M.O.

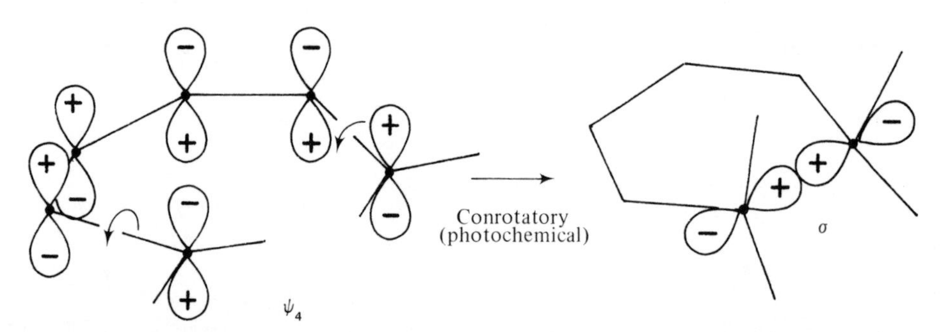

Q2.8 *cis,cis*-Octa-1,3,5,7-tetraene in its ground-state.

A2.8 Electronic configuration: $\psi_1^2\psi_2^2\psi_3^2\psi_4^2$
ψ_4 is the H.O.M.O.

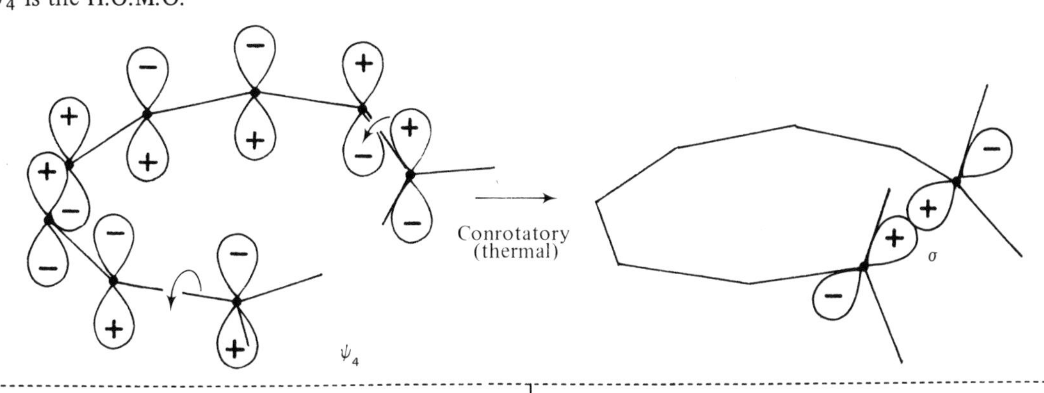

Q2.9 Prop-2-enyl cation in its ground-state.

A2.9 Electronic configuration: ψ_1^2
ψ_1 is the H.O.M.O.

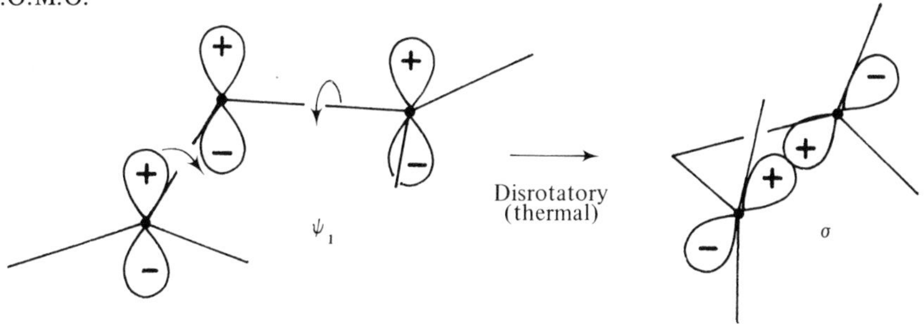

Q2.10 Prop-2-enyl anion in its ground-state.

A2.10 Electronic configuration: $\psi_1^2\psi_2^2$
ψ_2 is the H.O.M.O.

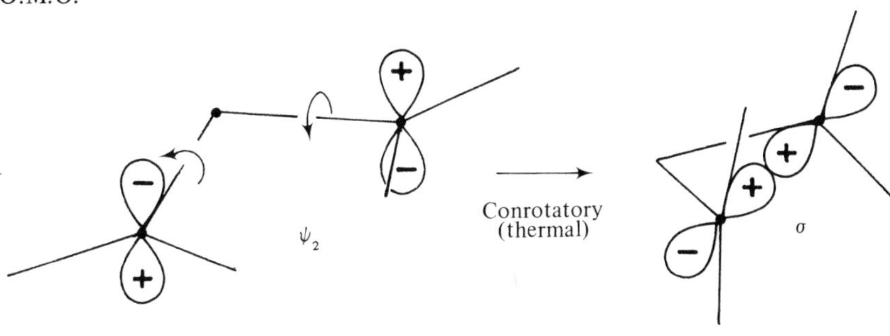

Q2.11 Penta-2,4-dienyl cation in its ground-state.

A2.11 Electronic configuration: $\psi_1^2\psi_2^2$
ψ_2 is the H.O.M.O.

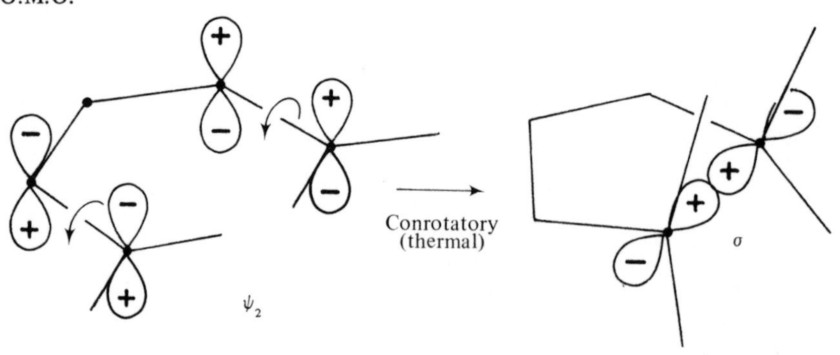

Q2.12 Penta-2,4-dienyl anion in its ground-state.

A2.12 Electronic configuration: $\psi_1^2\psi_2^2\psi_3^2$
ψ_3 is the H.O.M.O.

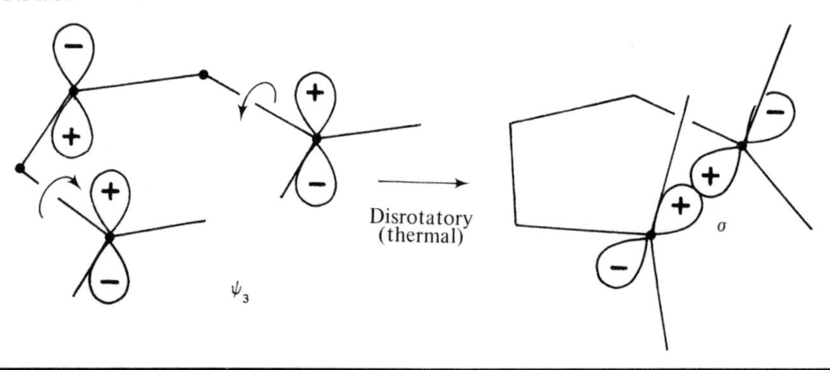

C2.2 Concerted, *thermal* cyclisation of linear π-systems will occur by a *conrotatory* mode if the linear π-system contains $4n$ π-electrons ($n = 1, 2, 3$, etc.), and by a *disrotatory* mode if the linear π-system contains $4n + 2$ π-electrons ($n = 0, 1, 2, 3$, etc.) (see Table below).

Concerted, *photochemical* cyclisation of linear π-systems from their 1st excited-state will occur by a *disrotatory* mode if $4n$ π-electrons are involved, and by a *conrotatory* mode if $4n + 2$ π-electrons are involved (see Table below).

TABLE

System	No. of π-electrons	Thermal reaction	Photochemical reaction
C=C—C$^+$	2	Disrotatory	Conrotatory
C=C—C$^-$	4	Conrotatory	Disrotatory
C=C—C=C	4	Conrotatory	Disrotatory
C=C—C=C—C$^+$	4	Conrotatory	Disrotatory
C=C—C=C—C$^-$	6	Disrotatory	Conrotatory
C=C—C=C—C=C	6	Disrotatory	Conrotatory
C=C—C=C—C=C—C$^+$	6	Disrotatory	Conrotatory
C=C—C=C—C=C—C$^-$	8	Conrotatory	Disrotatory
C=C—C=C—C=C—C=C	8	Conrotatory	Disrotatory
	$4n$ (n = 1, 2, 3, etc.)	Conrotatory	Disrotatory
	$4n + 2$ (n = 0, 1, 2, 3, etc.)	Disrotatory	Conrotatory

Q2.13 Which stereoisomer of 3,4-dimethylcyclobutene would you expect to be formed photochemically from *trans,trans*-hexa-2,4-diene?

A2.13

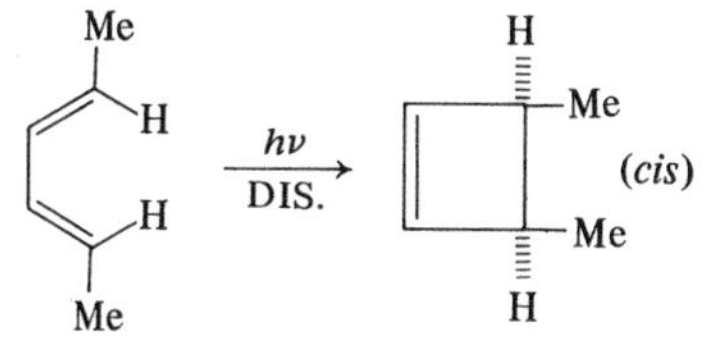

Me, H, H, Me; $h\nu$ DIS. → H, Me, Me, H (*cis*)

(R. Srinivasan, *J. Am. chem. Soc.*, 1968, **90**, 4498.)

Q2.14 Predict the stereochemistry of 5,6-dimethylcyclohexa-1,3-diene when formed (a) thermally, and (b) photochemically, from *trans,cis,trans*-octa-2,4,6-triene.

A2.14

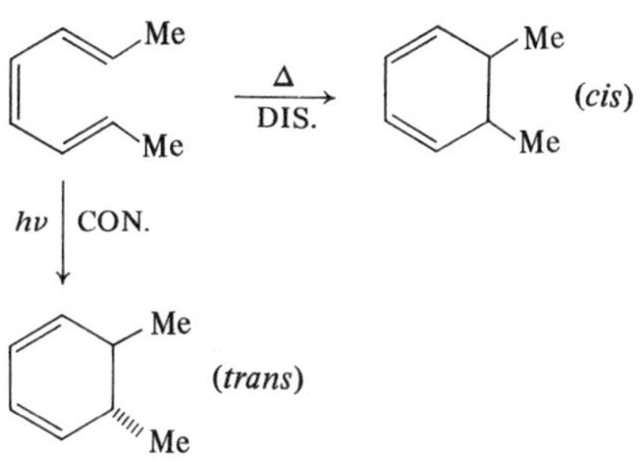

Me, Me; Δ DIS. → Me, Me (*cis*)

$h\nu$ CON. → Me, Me (*trans*)

[(a) E. N. Marvell, G. Caple and B. Schatz, *Tetrahedron Letters*, 1965, 385; E. Vogel, W. Grimme and E. Dinne, *ibid.*, 1965, 391. (b) G. J. Fonken, *ibid.*, 1962, 549.]

Q2.15 *cis,cis*-Cyclo-octa-1,3-diene, on treatment with (i) phenyl potassium, and (ii) H^+, gives bicyclo-[3, 3, 0]oct-2-ene. By assuming the initial formation of the cyclo-octa-2,4-dienyl anion, suggest a mechanism for this transformation. What stereochemistry would you predict for the product?

A2.15

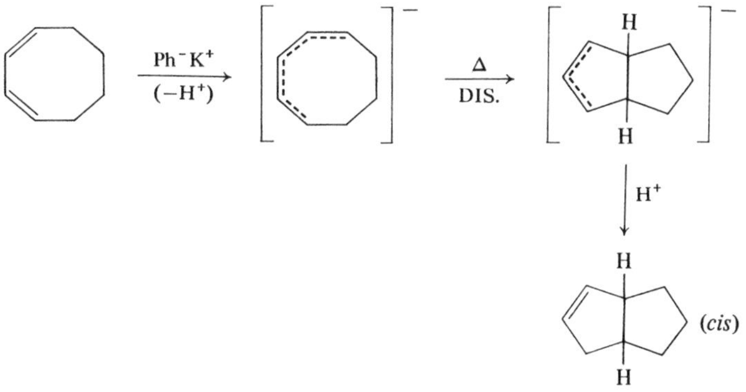

Ph^-K^+ $(-H^+)$ → [ ]$^-$ Δ DIS. → [H, H]$^-$ H^+ → H, H (*cis*)

(P. R. Stapp and R. F. Kleinschmidt, *J. org. Chem.*, 1965, **30**, 3006; R. B. Bates and D. A. McCombs, *Tetrahedron Letters*, 1969, 977.)

S2.3 We have only considered cyclisation reactions in this simplified approach, but electrocyclic reactions may also involve ring-opening. In the ring-opening direction, the stereochemical course will be the same as in the cyclisation direction, due to the principle of microscopic reversibility. For example, a concerted, thermal ring-opening of cyclobutene will occur by cleavage of the $C_{(3)}C_{(4)}$ σ-bond, with conrotatory rotation about the $C_{(1)}C_{(4)}$ and $C_{(2)}C_{(3)}$ bonds.

Orbital Correlation Diagrams

Introduction

In the Frontier Orbital approach we have only a superficial picture of the operation of the principle of conservation of orbital symmetry. What, for example, becomes of the remaining atomic orbitals when σ-bond formation occurs between the termini of the linear π-system? To answer questions of this type we need to consider all the molecular orbitals involved in the reaction and to follow their transformation throughout the chemical change.

S2.4 In order to determine if orbital symmetry is conserved during a reaction we have to examine the symmetry properties of all the molecular orbitals involved in the reaction with respect to the element of symmetry present in the geometrical change.

Q2.16 What element of symmetry is present in a *disrotatory* change?

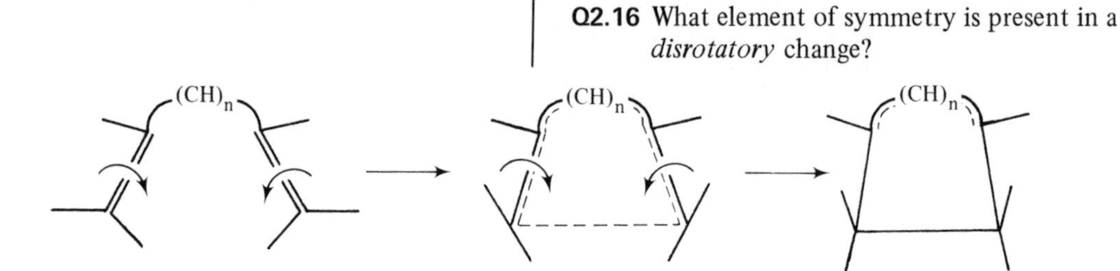

A2.16 A plane of symmetry, often written as m, is preserved throughout the change. This lies perpendicular to the plane of the linear π-system or the cyclised ring, and perpendicular to and bisecting the σ-bond which is formed or broken during the change. At any stage in the reaction, the image of one half of the system produced by reflection in the plane of symmetry is identical to the other half of the system.

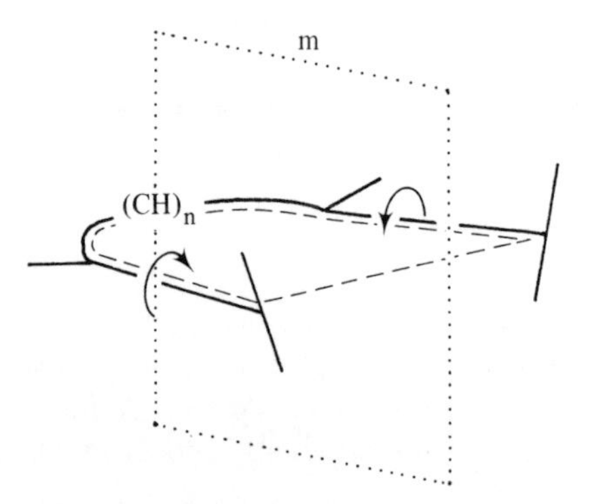

Q2.17 What element of symmetry is present in a *conrotatory* change?

A2.17 A 2-fold axis of symmetry, often written as C_2, C denoting an axis of symmetry with the subscript indicating the order of the rotation axis, is preserved throughout the change. This lies in the plane of the linear π-system or the cyclised ring, and perpendicular to and bisecting the σ-bond which is formed or broken during the change. At any stage in the reaction, rotation of the whole system through 180° about the axis of symmetry gives an arrangement which is indistinguishable from the original system.

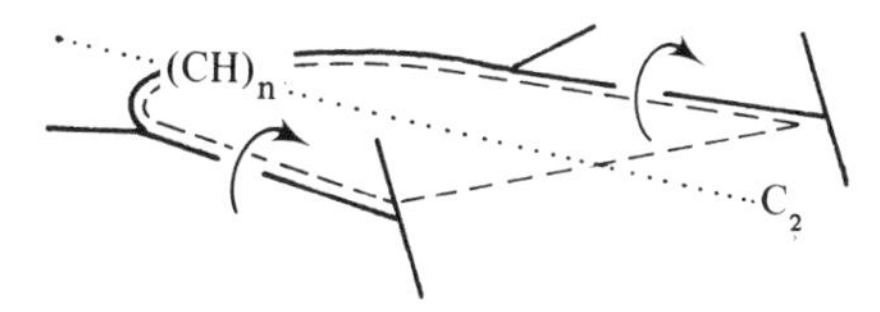

S2.5 To classify the symmetry of a molecular orbital for a particular element of symmetry, we perform the given symmetry operation on the molecular orbital. If the operation produces no change, then the orbital is classified as *symmetric.* If the operation produces a phase-inversion, then the orbital is classified as *anti-symmetric.* Consider ψ_2 of buta-1,3-diene:

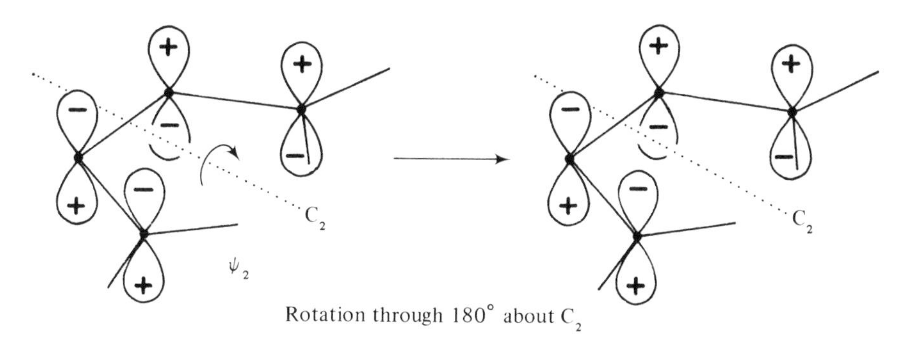

Rotation through 180° about C_2

With respect to the 2-fold axis of rotation shown, this orbital is *symmetric*, since after rotation through 180° the new orbital is indistinguishable from the original.

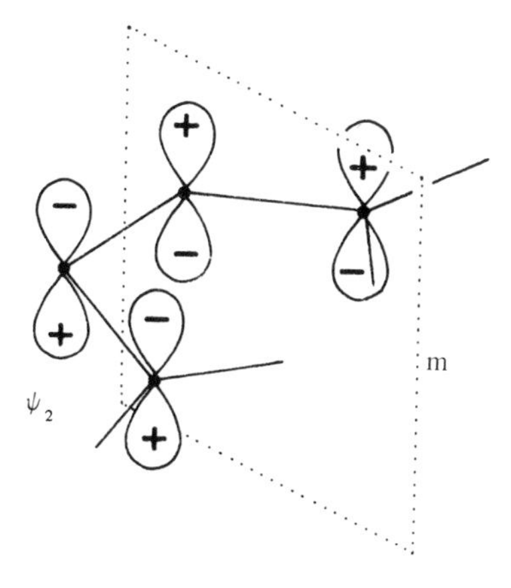

With respect to the plane of symmetry shown, this orbital is *antisymmetric*, since reflection of one half of the orbital in the plane of symmetry does not produce the other half, but a phase-inverted form of it.

Q2.18 to 2.23 Classify the following molecular orbitals as *symmetric* or *antisymmetric* with respect to the symmetry element shown:

Q2.18

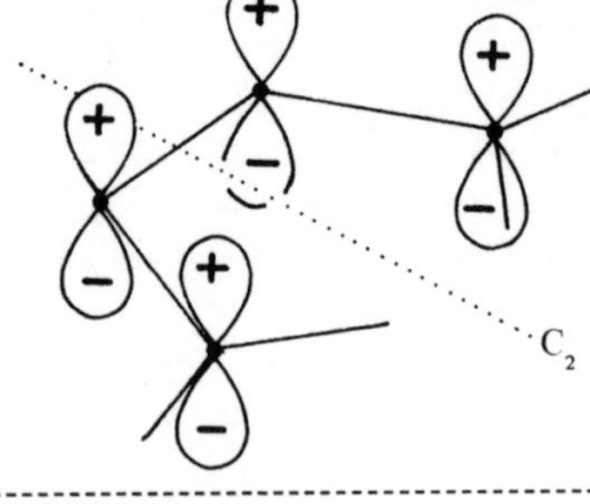

ψ_1 of buta-1,3-diene

C_2

A2.18 Antisymmetric

Q2.19

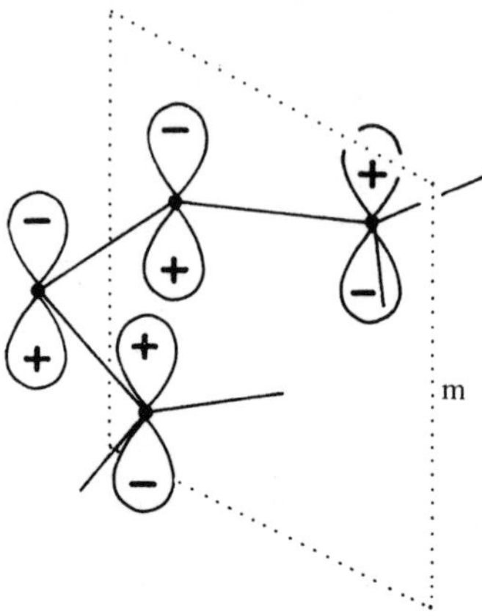

ψ_3 of buta-1,3-diene

m

A2.19 Symmetric.

Q2.20

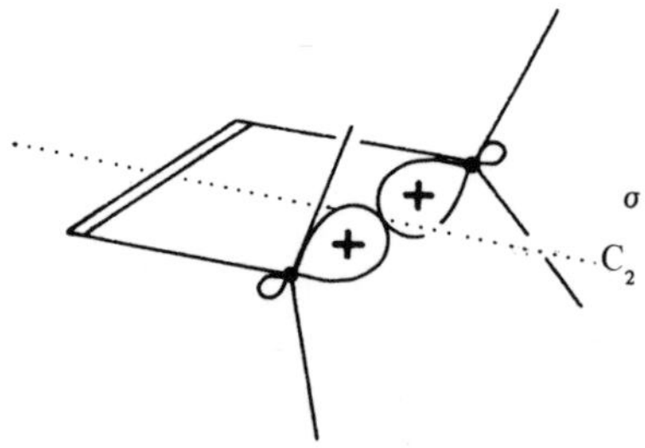

σ of cyclobutene

C_2

A2.20 Symmetric.

Q2.21

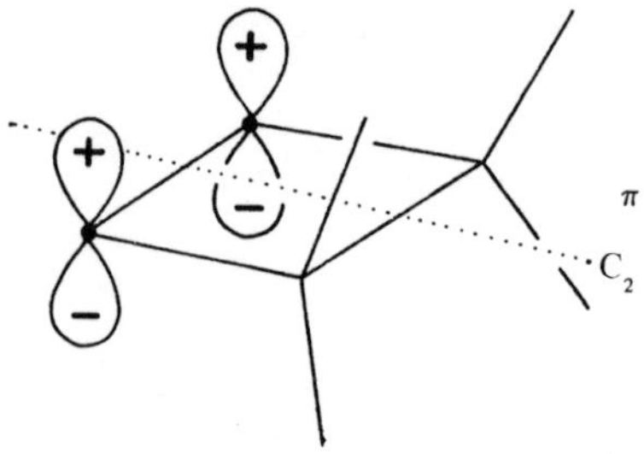

π of cyclobutene

C_2

A2.21 Antisymmetric.

Q2.22

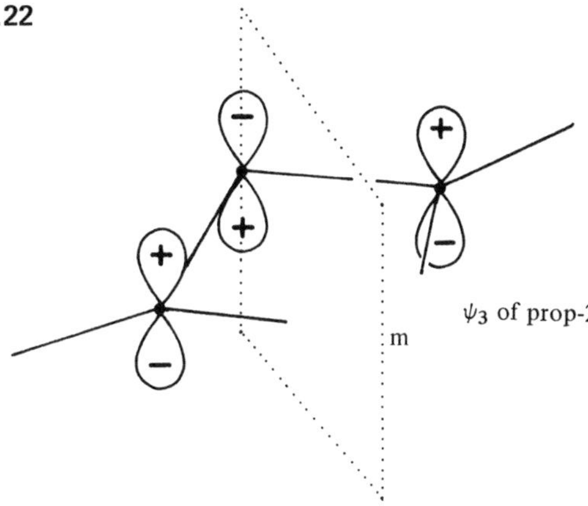

ψ_3 of prop-2-enyl

A2.22 Symmetric.

Q2.23

ψ_5 of penta-2,4-dienyl

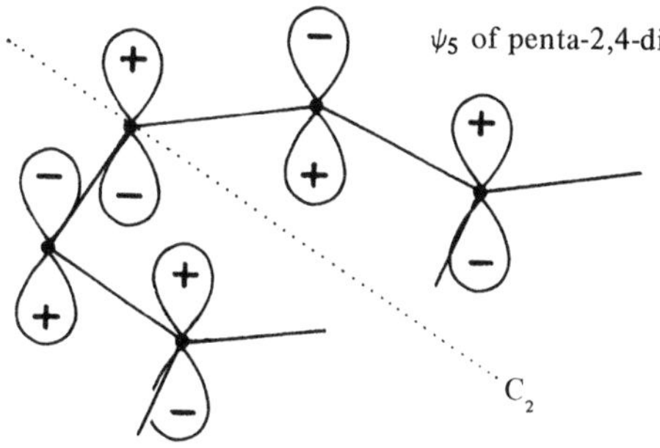

A2.23 Antisymmetric.

S2.6 During an electrocyclic reaction, those molecular orbitals of the reactant which are classified as symmetric with respect to the element of symmetry present in the change 'mix' together and emerge at the end of the reaction as symmetric molecular orbitals of the product. Similarly, the antisymmetric orbitals of the reactant 'mix' together and emerge as antisymmetric orbitals of the product.

When two orbitals of different energy but having the same symmetry classification mix together, the orbital of lower energy mixes into itself the orbital of higher energy in a bonding manner. Conversely, the orbital of higher energy mixes into itself the orbital of lower energy in an antibonding manner.

[These rules express the outcome of detailed molecular orbital calculations performed on the reactant, product, and many species of intermediate geometry encountered on passing through the transition state of a concerted reaction.]

Q2.24 For the buta-1,3-diene-cyclobutene transformation by a conrotatory rotation, which are the symmetric molecular orbitals of the reactant and product which are involved in the reaction?

A2.24 Conrotatory rotation–2-fold axis of symmetry. The symmetric orbitals of butadiene are ψ_2 and ψ_4, and those of cyclobutene are σ and π^*.

S2.7 Consider first the ring-opening reaction of cyclobutene: σ (lower energy), after cleavage and conrotatory rotation, mixes in π^* (higher energy) in a bonding manner.

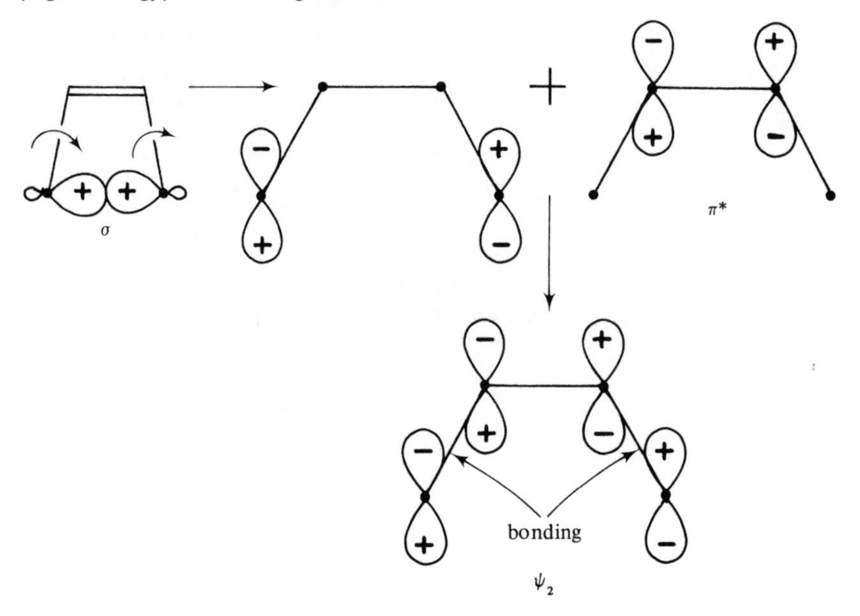

Thus, $\sigma + \pi^* \rightarrow \psi_2$.

At the same time π^* mixes in σ (after cleavage and conrotatory rotation) in an antibonding manner.

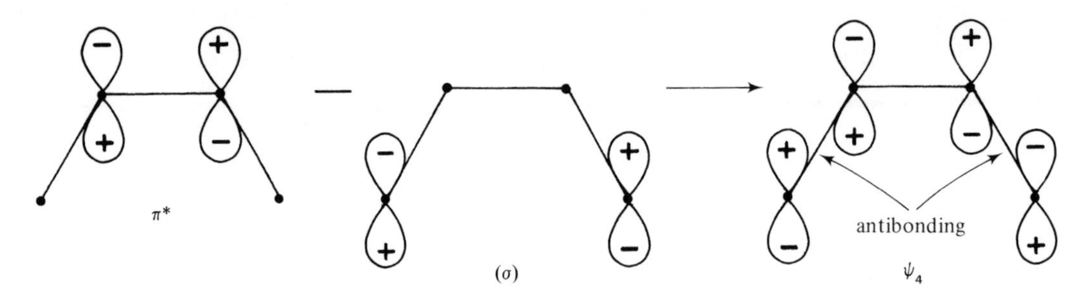

Thus, $\pi^* - \sigma \rightarrow \psi_4$.

Similarly, for the ring-closure reaction:

ψ_2 (after conrotatory rotation) mixes in ψ_4 (also after conrotatory rotation) in a bonding manner.

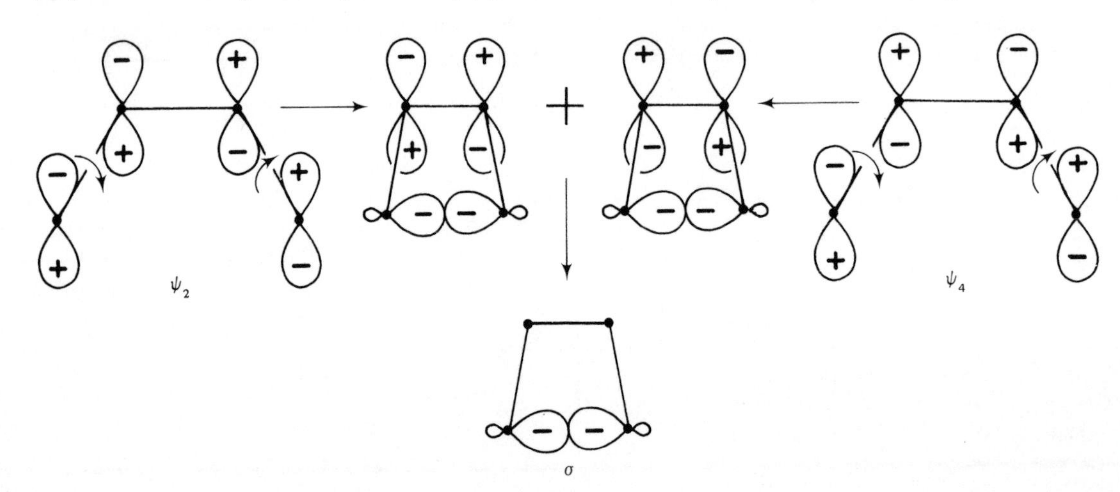

Thus, $\psi_2 + \psi_4 \rightarrow \sigma$.

At the same time, ψ_4 mixes in ψ_2 (both after conrotatory rotation) in an antibonding manner.

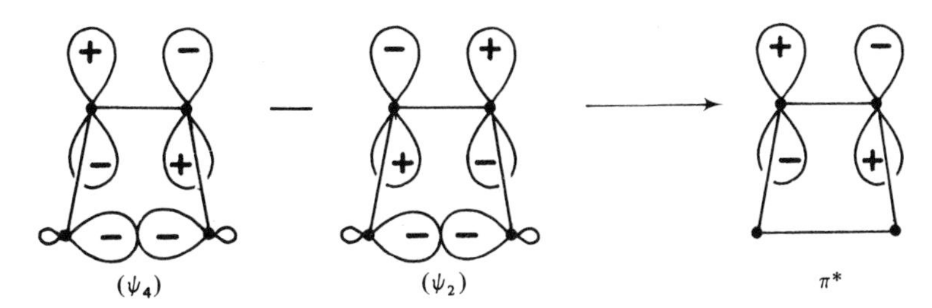

Thus, $\psi_4 - \psi_2 \rightarrow \pi^*$.

Of course, these changes do not occur as discrete steps, but as a continuous change, i.e. for ring-opening, bond cleavage, rotation and mixing occur simultaneously, but for analytical purposes it is useful to treat it in this manner.

(*N.B.* The pictorial representations given here, and also required in answer to Questions 2-26 to 2-29 and 2-32 to 2-35, are intended as a *rationalisation* rather than a *proof* of the orbital mixing. The orbitals produced by the mixing may be deduced by the application of Statement 2.6 (page 21), and should be known before a pictorial representation is attempted.)

Q2.25 For the buta-1,3-diene-cyclobutene transformation by a conrotatory rotation, which are the antisymmetric molecular orbitals of the reactant and product which are involved in the reaction?

A2.25 Conrotatory rotation—C_2.
The antisymmetric orbitals of butadiene are ψ_1 and ψ_3, and those of cyclobutene are π and σ^*.

Q2.26 For ring-opening by a conrotatory rotation, π mixes in σ^* (after cleavage and conrotatory rotation) in a bonding manner. Does this combination give ψ_1 or ψ_3? Draw a pictorial representation of this, as shown previously for the symmetric orbitals.

A2.26 $\pi + \sigma^* \rightarrow \psi_1$

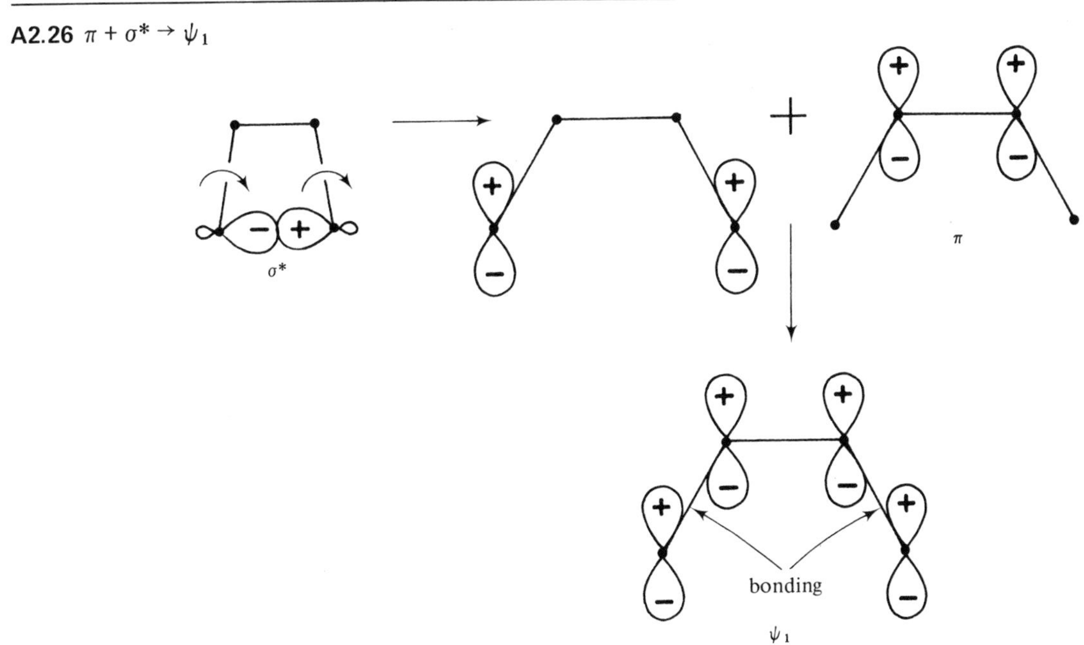

Q2.27 Draw a pictorial representation of the $\sigma^* - \pi$ combination. Which orbital of butadiene does this produce?

A2.27 $\sigma^* - \pi \rightarrow \psi_3$

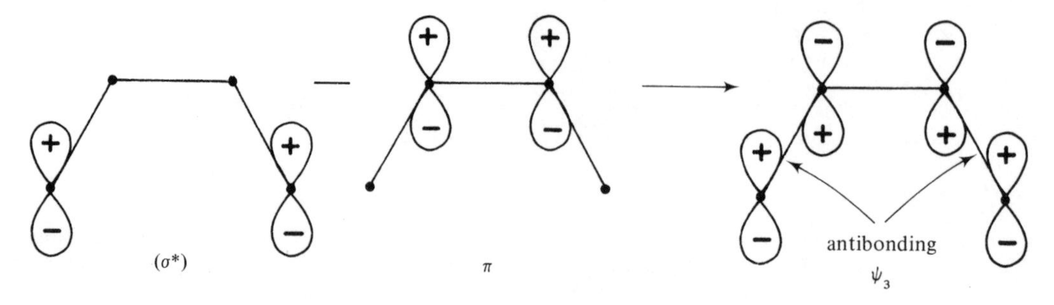

Q2.28 For ring-closure by a conrotatory rotation, ψ_1 mixes in ψ_3 (both after conrotatory rotation) in a bonding manner. Does this combination give π or σ^*? Draw a pictorial representation of this.

A2.28 $\psi_1 + \psi_3 \rightarrow \pi$

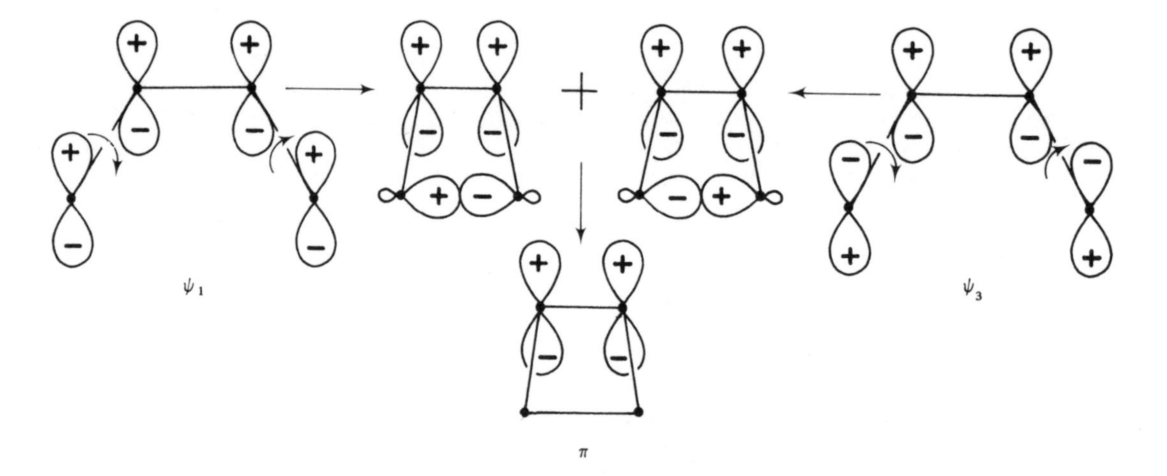

Q2.29 Draw a pictorial representation of the $\psi_3 - \psi_1$ combination. Which orbital of cyclobutene does this produce?

A2.29 $\psi_3 - \psi_1 \rightarrow \sigma^*$

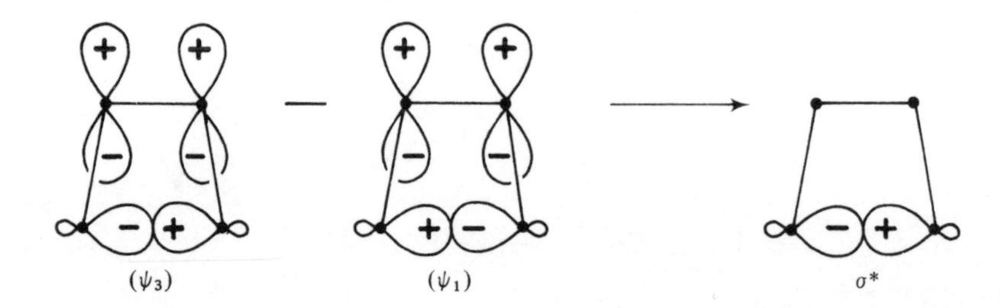

Q2.30 For the conrotatory ring-opening of cyclobutene, pictorially represent the hypothetical mixing of σ (symmetric, after conrotatory rotation) and π (antisymmetric). Does this give acceptable molecular orbitals of butadiene?

A2.30

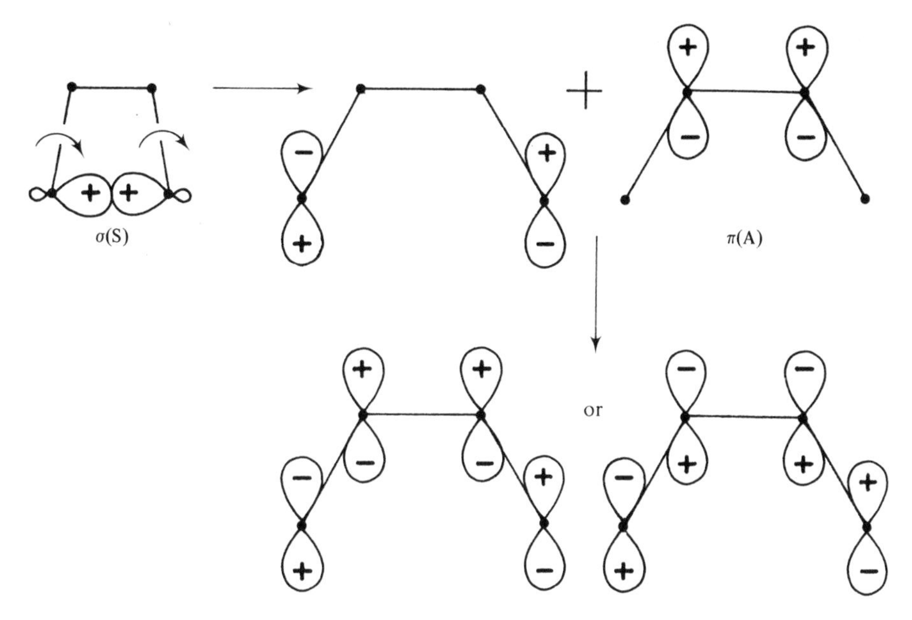

These are not acceptable molecular orbitals of butadiene since they are *unsymmetrical.* Orbitals must have the *same* symmetry classification in order to 'mix' properly.

Q2.31 For the buta-1,3-diene-cyclobutene transformation by a disrotatory rotation, classify the molecular orbitals of the reactant and product which are involved in the reaction as either symmetric or antisymmetric.

A2.31 Disrotatory rotation–plane of symmetry (m)

Butadiene: ψ_1 and ψ_3 are symmetric
ψ_2 and ψ_4 are antisymmetric
Cyclobutene: σ and π are symmetric
π^* and σ^* are antisymmetric

Q2.32 For ring-opening by a disrotatory rotation, state which combinations of σ and π lead to ψ_1 and ψ_3 respectively (all symmetric). Represent these combinations pictorially.

A2.32 $\sigma + \pi \rightarrow \psi_1$
$\pi - \sigma \rightarrow \psi_3$

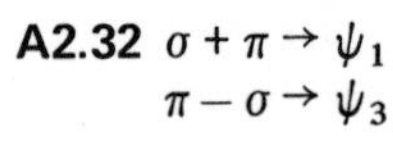

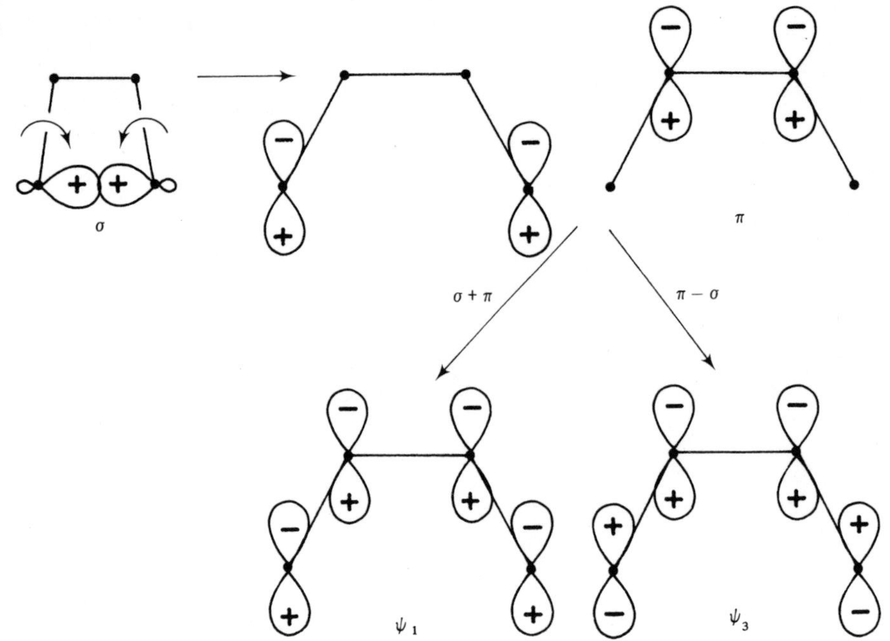

Q2.33 For ring-closure by a disrotatory rotation, state which combinations of ψ_1 and ψ_3 lead to σ and π respectively. Represent these combinations pictorially.

A2.33 $\psi_1 + \psi_3 \rightarrow \sigma$ (product orbital of lower energy)
$\psi_3 - \psi_1 \rightarrow \pi$ (product orbital of higher energy)

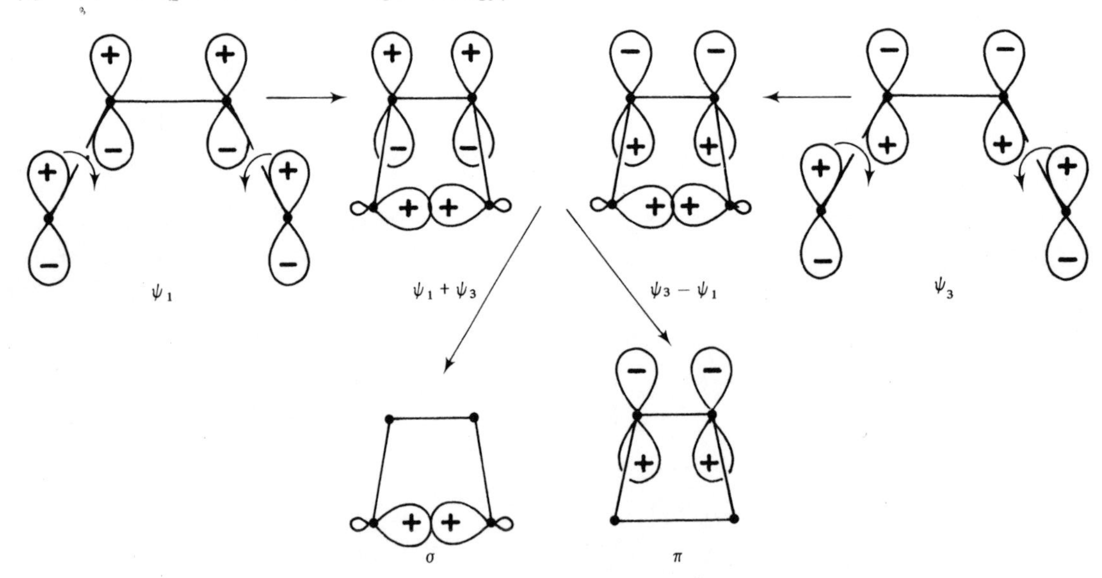

Q2.34 For ring-opening by a disrotatory rotation, state which combinations of π^* and σ^* lead to ψ_2 and ψ_4 respectively (all antisymmetric). Represent these combinations pictorially.

A2.34 $\pi^* + \sigma^* \rightarrow \psi_2$
$\sigma^* - \pi^* \rightarrow \psi_4$

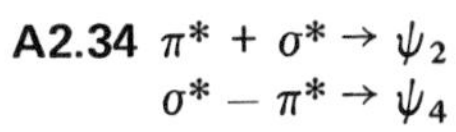

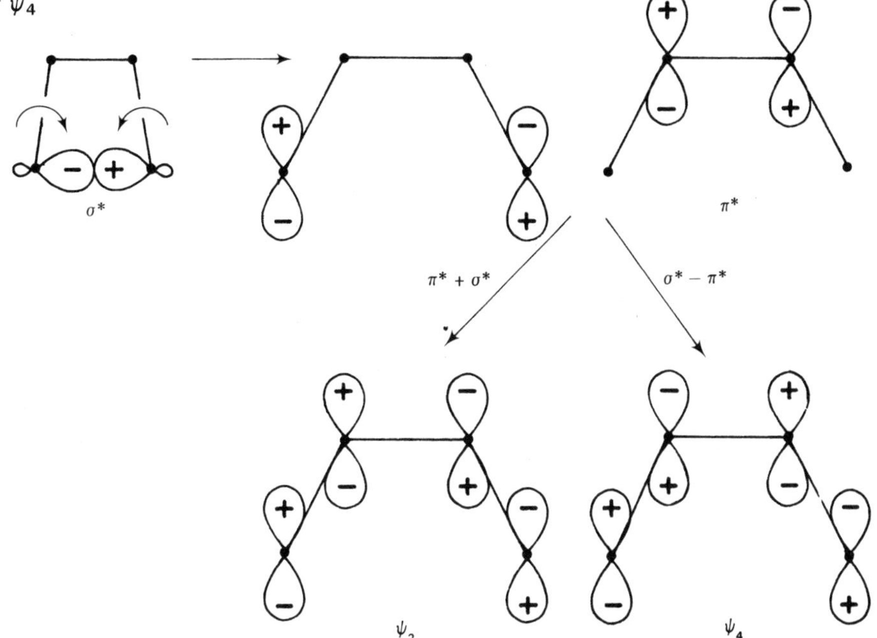

Q2.35 For ring-closure by a disrotatory rotation, state which combinations of ψ_2 and ψ_4 lead to π^* and σ^* respectively. Represent these combinations pictorially.

A2.35 $\psi_2 + \psi_4 \rightarrow \pi^*$ (product orbital of lower energy)
$\psi_4 - \psi_2 \rightarrow \sigma^*$ (product orbital of higher energy)

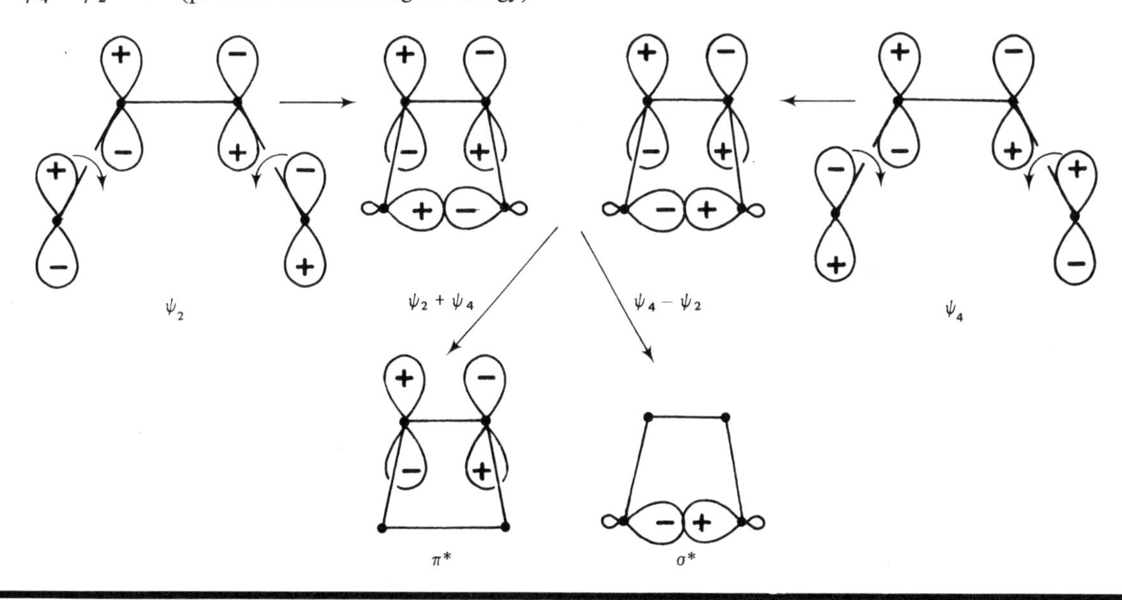

S2.8 The results obtained from Statement 2.7 and Questions 2.26 to 2.29 enable us to draw up the following orbital correlation table for the buta-1,3-diene-cyclobutene transformation by a *conrotatory* rotation:

Symmetric orbitals	*Antisymmetric orbitals*
$\psi_2 + \psi_4 \rightarrow \sigma$	$\psi_1 + \psi_3 \rightarrow \pi$
$\psi_4 - \psi_2 \rightarrow \pi^*$	$\psi_3 - \psi_1 \rightarrow \sigma^*$
$\sigma + \pi^* \rightarrow \psi_2$	$\pi + \sigma^* \rightarrow \psi_1$
$\pi^* - \sigma \rightarrow \psi_4$	$\sigma^* - \pi \rightarrow \psi_3$

With this information we can construct an orbital correlation diagram:

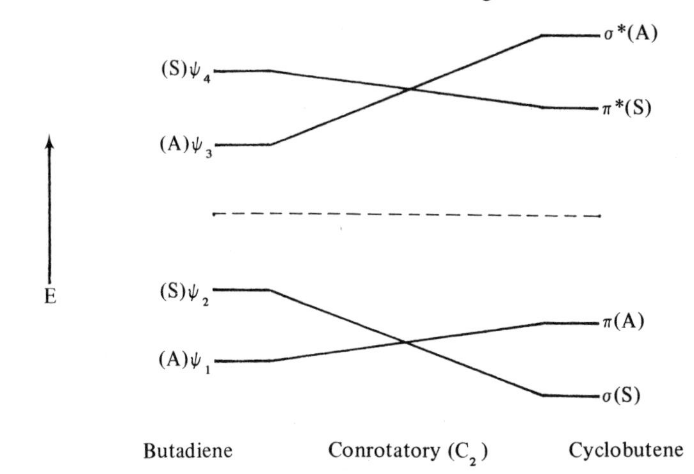

To do this we place the orbitals in order of increasing energy—the orbital of lowest energy at the bottom—and indicate their symmetry classification with respect to the element of symmetry present in the transformation (S, symmetric; A, antisymmetric). We then draw lines (correlations) connecting orbitals of the same symmetry classification, as shown. Lines connecting *two pairs* of orbitals of the *same symmetry* must not cross; e.g. ψ_2 does not correlate with π^*, or ψ_4 with σ. [This last condition follows from S2.6 (page 21) concerning mixing of orbitals of the same symmetry.]

Notice that the diagram contains all the information regarding orbital correlations which we have just tabulated. To obtain an orbital correlation diagram we do not *have* to work through a detailed analysis of orbital mixing. If we simply follow the rules just stated, we will obtain the correct orbital correlation diagram. We shall presently see how useful these diagrams are.

Q2.36 Using the results of Questions 2.32 to 2.35 tabulate the orbital correlations for the buta-1,3-diene-cyclobutene transformation by a disrotatory rotation.

A2.36

Symmetric orbitals	*Antisymmetric orbitals*
$\psi_1 + \psi_3 \rightarrow \sigma$	$\psi_2 + \psi_4 \rightarrow \pi^*$
$\psi_3 - \psi_1 \rightarrow \pi$	$\psi_4 - \psi_2 \rightarrow \sigma^*$
$\sigma + \pi \rightarrow \psi_1$	$\pi^* + \sigma^* \rightarrow \psi_2$
$\pi - \sigma \rightarrow \psi_3$	$\sigma^* - \pi^* \rightarrow \psi_4$

Q2.37 Arrange the orbitals of buta-1,3-diene, and of cyclobutene, in order of increasing energy.

A2.37 (↑ E)

Butadiene	Cyclobutene
ψ_4	σ^*
ψ_3	π^*
ψ_2	π
ψ_1	σ

Q2.38 Classify the symmetry of each orbital of buta-1,3-diene and cyclobutene with respect to the plane of symmetry present in a disrotatory rotation.

A2.38 (A) ψ_4 σ^* (A)
(S) ψ_3 π^* (A)
(A) ψ_2 π (S)
(S) ψ_1 σ (S)

Q2.39 Construct an orbital correlation diagram for the disrotatory interconversion of buta-1,3-diene and cyclobutene.

A2.39

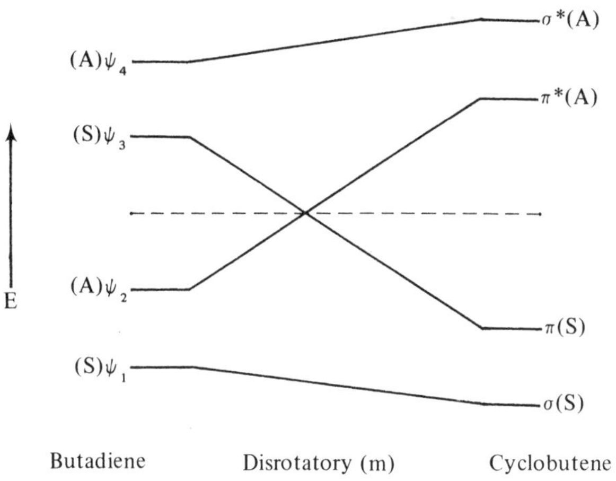

σ^*(A)
(A)ψ_4
π^*(A)
(S)ψ_3
(A)ψ_2
E
π(S)
(S)ψ_1
σ(S)
Butadiene Disrotatory (m) Cyclobutene

Q2.40 What is the ground-state (G.S.) electronic configuration of butadiene?

A2.40 $\psi_1^2\psi_2^2$

Q2.41 Using the orbital correlation diagram for a *conrotatory* change, predict the electronic configuration of cyclobutene derived from ground-state butadiene by this mode of rotation.

A2.41 $\psi_1^2 \rightarrow \pi^2$, $\psi_2^2 \rightarrow \sigma^2$
$\therefore \psi_1^2\psi_2^2$ (G.S.) $\rightarrow \sigma^2\pi^2$ (G.S.)

Q2.42 Using the orbital correlation diagram for a *disrotatory* change, predict the electronic configuration of cyclobutene derived from ground-state butadiene by this mode of rotation.

A2.42 $\psi_1^2 \rightarrow \sigma^2$, $\psi_2^2 \rightarrow \pi^{*2}$
$\therefore \psi_1^2\psi_2^2$ (G.S.) $\rightarrow \sigma^2\pi^{*2}$ (an excited-state of cyclobutene)

Q2.43 For the conversion of butadiene into cyclobutene by a concerted thermal reaction, which mode of rotation would require the least activation energy from conservation of orbital symmetry considerations?

A2.43 Conrotatory rotation gives cyclobutene in its ground-state electronic configuration, whereas disrotatory rotation gives cyclobutene in an excited-state electronic configuration. Therefore conrotatory rotation will require less energy ($E_{\sigma^2\pi^{*2}} > E_{\sigma^2\pi^2}$).

Q2.44 For the conversion of cyclobutene into butadiene by a concerted, thermal reaction, which mode of rotation would require the least activation energy from conservation of orbital symmetry considerations?

A2.44 Conrotatory rotation: $\sigma^2\pi^2$ (G.S.) $\rightarrow \psi_1^2\psi_2^2$ (G.S.)
Disrotatory rotation: $\sigma^2\pi^2$ (G.S.) $\rightarrow \psi_1^2\psi_3^2$ (an excited-state electronic configuration).

Therefore conrotatory rotation would require the least energy ($E_{\psi_1^2\psi_3^2} > E_{\psi_1^2\psi_2^2}$).

Q2.45 What is the electronic configuration of the 1st excited-state of butadiene?

A2.45 $\psi_1^2\psi_2^1\psi_3^1$ ($\psi_2 \xrightarrow{h\nu} \psi_3$).

Q2.46 Using the orbital correlation diagrams, predict the electronic configuration of cyclobutene derived from butadiene in its 1st excited-state by (a) conrotatory, and (b) disrotatory, rotation.

A2.46 (a) Conrotatory: $\psi_1^2 \rightarrow \pi^2$, $\psi_2^1 \rightarrow \sigma^1$, $\psi_3^1 \rightarrow \sigma^{*1}$
$\therefore \psi_1^2 \psi_2^1 \psi_3^1$ (1st excited-state) $\rightarrow \sigma^1 \pi^2 \sigma^{*1}$ (highly excited-state of cyclobutene).

(b) Disrotatory: $\psi_1^2 \rightarrow \sigma^2$, $\psi_2^1 \rightarrow \pi^{*1}$, $\psi_3^1 \rightarrow \pi^1$
$\therefore \psi_1^2 \psi_2^1 \psi_3^1$ (1st excited-state) $\rightarrow \sigma^2 \pi^1 \pi^{*1}$ (1st excited-state of cyclobutene).

Q2.47 For the conversion of butadiene into cyclobutene by a concerted, photochemical reaction (from the 1st excited-state of butadiene), which mode of rotation would require the least activation energy from conservation of orbital symmetry considerations?

A2.47 Conrotatory rotation: $\psi_1^2 \psi_2^1 \psi_3^1 \rightarrow \sigma^1 \pi^2 \sigma^{*1}$
Disrotatory rotation: $\psi_1^2 \psi_2^1 \psi_3^1 \rightarrow \sigma^2 \pi^1 \pi^{*1}$
Disrotatory rotation would require the least energy ($E_{\sigma^1 \pi^2 \sigma^{*1}} > E_{\sigma^2 \pi^1 \pi^{*1}}$).

C2.3 Thermal interconversion of butadiene and cyclobutene will occur *via* conrotatory rotation, and photochemical interconversion *via* disrotatory rotation. (The simple Frontier Orbital treatment gave the same predictions.)

Q2.48 Which geometrical isomer of hexa-2,4-diene would you predict to be formed by thermal cleavage of *cis*-3,4-dimethylcyclobutene?

A2.48

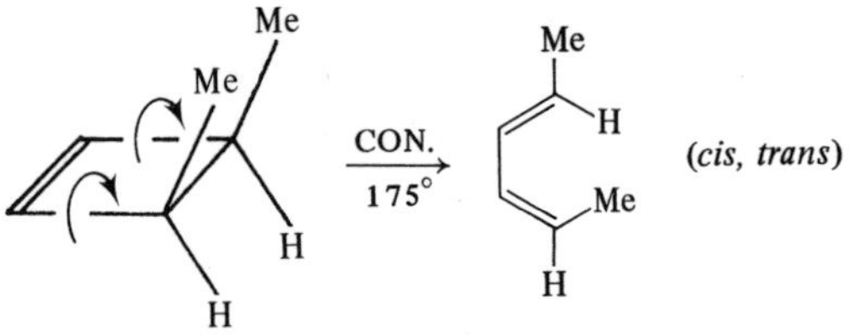

CON., 175° → (*cis, trans*)

(R. E. K. Winter, *Tetrahedron Letters,* 1965, 1207; J. I. Brauman and W. C. Archie, *J. Am. chem. Soc.,* 1972, **94**, 4262.)

Q2.49 By considering the two possible conrotatory modes of cleavage of *trans*-3,4-dimethylcyclobutene, predict which isomer of hexa-2,4-diene would be formed in a thermal reaction. Why should only one isomer be formed?

A2.49 (a)

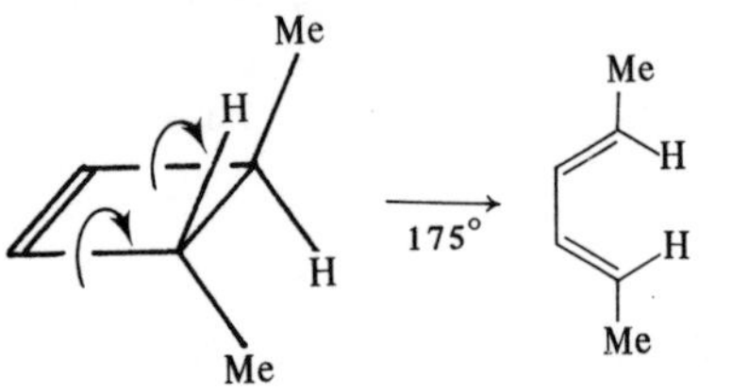

175° →

(b)

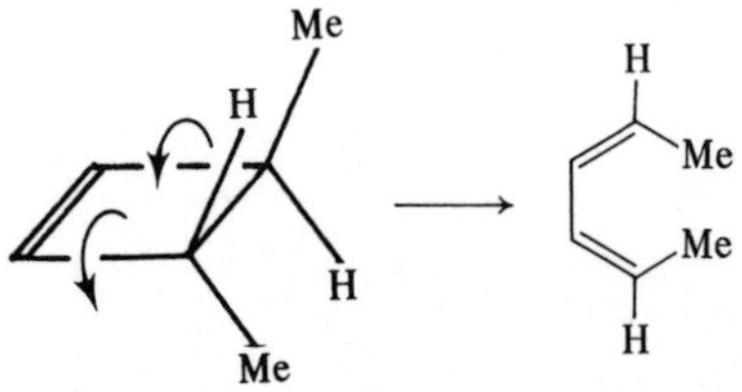

(cont.)

The steric interactions (methyl, methyl) present in the transition state for reaction (b) will be greater than those (hydrogen, hydrogen) present in the transition state for reaction (a). Therefore, reaction (a) will be favoured, giving *trans,trans*-hexa-2,4-diene.

(R. E. K. Winter, *Tetrahedron Letters*, 1965, 1207.)

Q2.50 The thermal conversion of *A* into *B* has a half-life of 2 hr. at 185°, whereas the systems *C, D,* and *E* require reaction temperatures of 350°, 335°, and 180° respectively, for similar electrocyclic reactions to occur at the same rate. Rationalise these results.

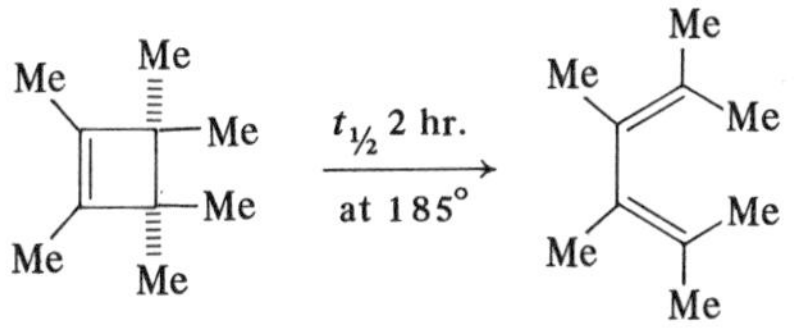

Me Me Me Me Me Me (*A*) $\xrightarrow[\text{at } 185^\circ]{t_{1/2}\ 2\text{ hr.}}$ Me Me Me Me Me Me (*B*)

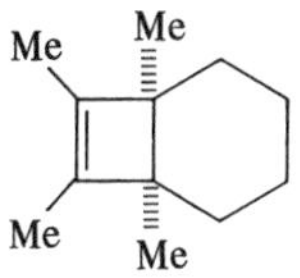

(*C*; $t_{1/2}$ 2 hr. at 350°) 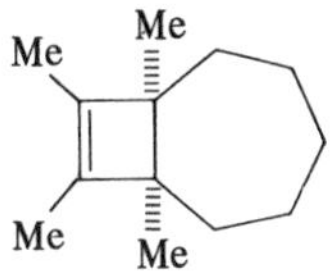 (*D*; $t_{1/2}$ 2 hr. at 335°) 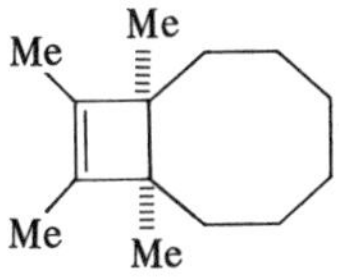 (*E*; $t_{1/2}$ 2 hr. at 180°)

A2.50 Reaction $A \rightarrow B$ presumably occurs by a conrotatory mode. Electrocyclic reactions of *C* and *D* by the same mode would increase the angle strain in these systems since the intended product would be a *cis,trans*-1,3-diene contained in an 8- and a 9-membered ring respectively. (Reaction of *C* and *D* may in fact occur by a non-concerted, disrotatory path.) The conrotatory reaction of *E* can occur without an appreciable increase in angle strain, and the product is probably *cis,trans*-1,2,3,4-tetramethylcyclodeca-1,3-diene.

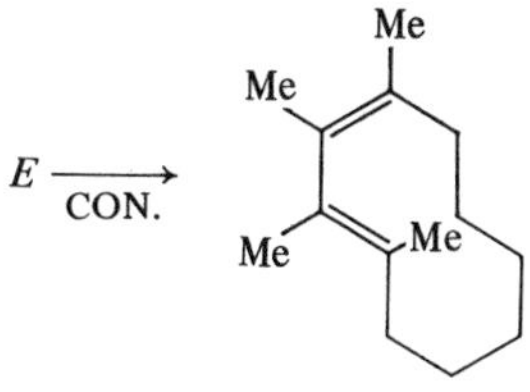

$E \xrightarrow{\text{CON.}}$ Me Me Me Me

(R. Criegee, *Angew. Chem. Int. Ed.*, 1968, 7, 559.)

Q2.51 What would you predict the stereochemistry of bicyclo[4, 2, 0]oct-7-ene to be, when formed by a photochemical, electrocyclic reaction of *cis,cis*-cyclo-octa-1,3-diene? What would be the stereochemistry of the same product if the reaction occurred in two steps: (i) initial photoisomerisation to *cis, trans*-cyclo-octa-1,3-diene, followed by (ii) a thermal electrocyclisation?

A2.51

The same product, *cis*-bicyclo[4, 2, 0]oct-7-ene would be formed by both routes.

(R. S. H. Liu, *J. Am. chem. Soc.*, 1967, **89**, 112 and references therein.)

Q2.52 For the prop-2-enyl-cyclopropyl interconversion, the molecular orbitals which are relevant to an analysis for conservation of orbital symmetry are:

prop-2-enyl

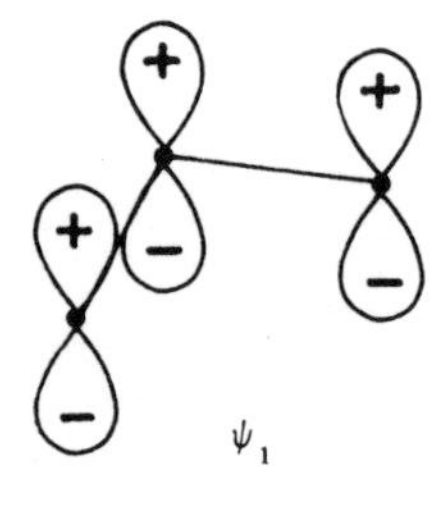

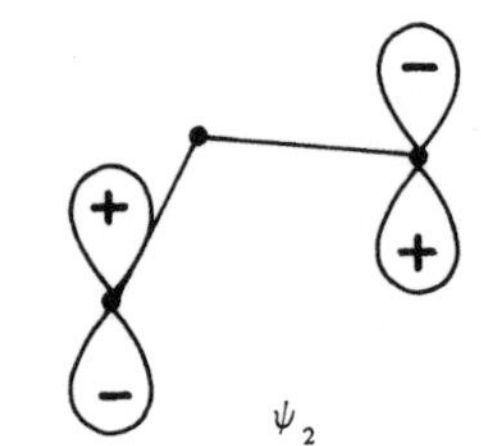

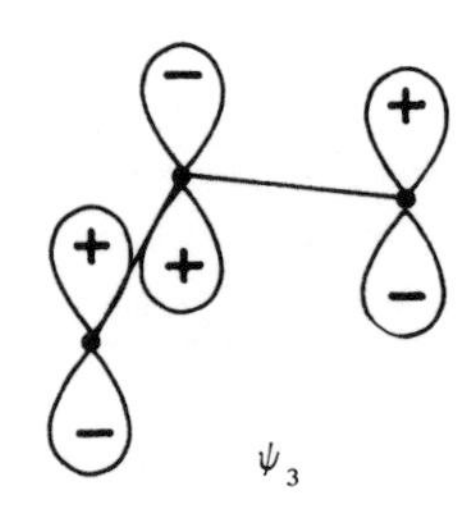

cyclopropyl

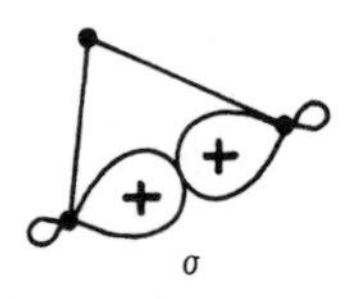

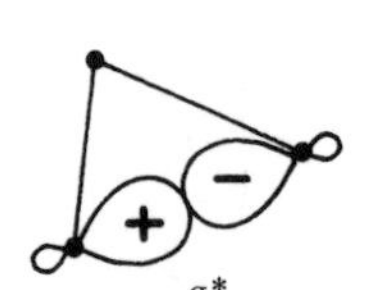

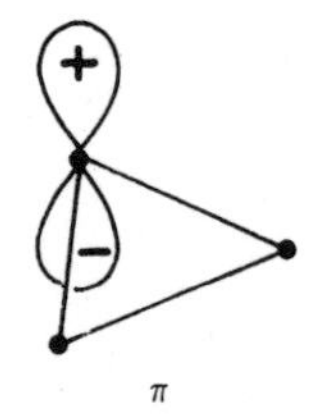

Classify the symmetry of these orbitals for both conrotatory and disrotatory modes of reaction.

A2.52 Conrotatory rotation (C_2)

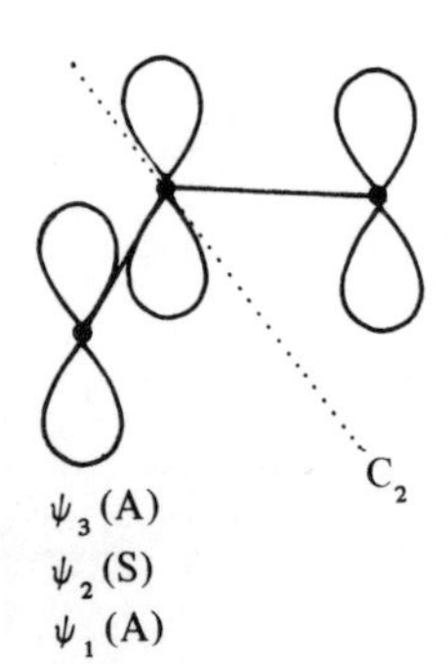

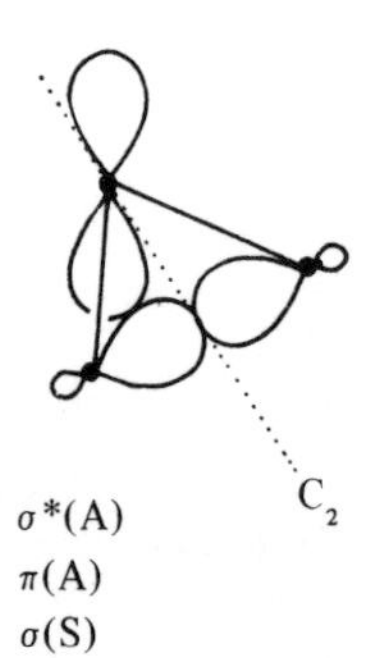

(cont.)

Disrotatory rotation (m)

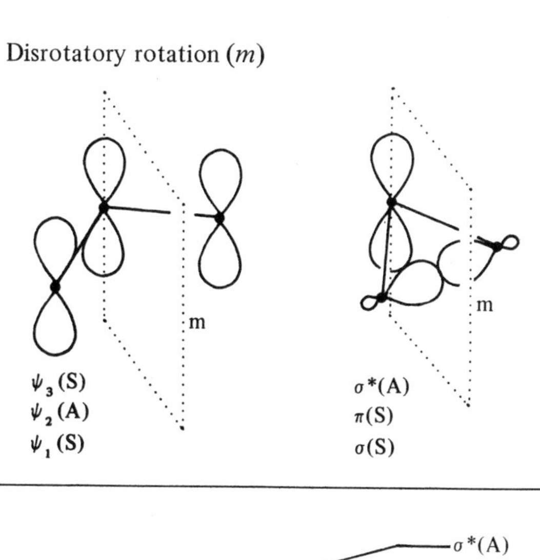

Q2.53 Construct an orbital correlation diagram for the conrotatory interconversion of prop-2-enyl and cyclopropyl species.

A2.53

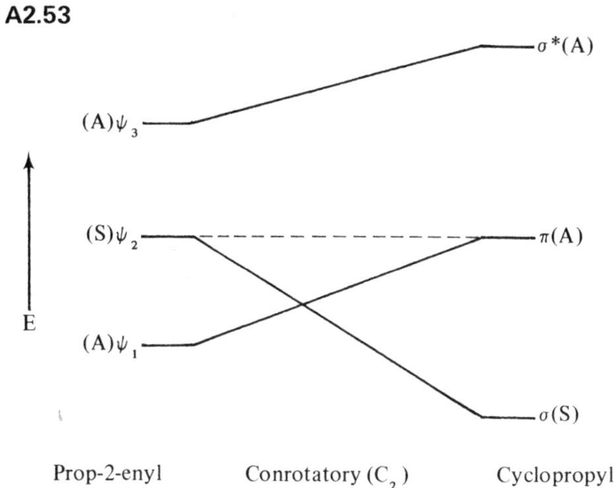

Q2.54 Construct an orbital correlation diagram for the disrotatory interconversion of prop-2-enyl and cyclopropyl species.

A2.54

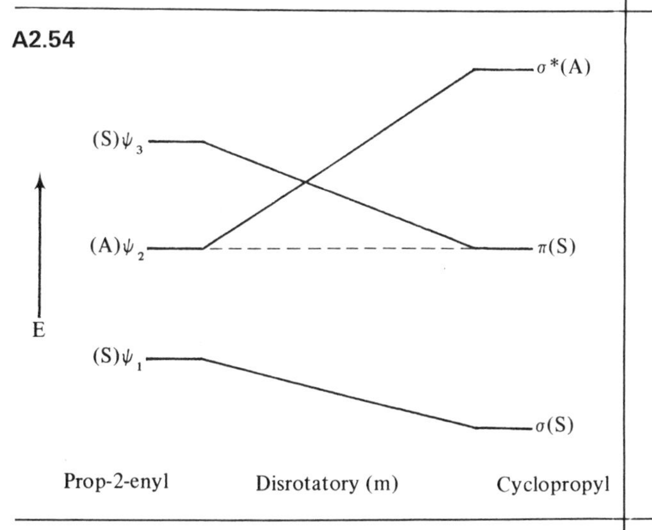

Q2.55 What is the electronic configuration of the cyclopropyl cation in its ground-state?

A2.55 σ^2

Q2.56 Using the orbital correlation diagrams, predict the electronic configuration of a prop-2-enyl cation derived from a cyclopropyl cation in its ground-state by (a) conrotatory, and (b) disrotatory, rotation.

A2.56 (a) Conrotatory: σ^2 (G.S.) $\rightarrow \psi_2^2$ (an excited-state).
(b) Disrotatory: σ^2 (G.S.) $\rightarrow \psi_1^2$ (G.S.)

Q2.57 For the conversion of a cyclopropyl cation into a prop-2-enyl cation by a concerted, thermal reaction, which mode of rotation would require the least activation energy from conservation of orbital symmetry considerations?

A2.57 Disrotatory rotation would require the least energy ($E_{\psi_2^2} > E_{\psi_1^2}$).

Q2.58 What is the electronic configuration of the 1st excited-state of the prop-2-enyl anion?

A2.58 $\psi_1^2\psi_2^1\psi_3^1$ ($\psi_2 \xrightarrow{h\nu} \psi_3$)

Q2.59 Using the orbital correlation diagrams, predict the electronic configuration of a cyclopropyl anion derived from a prop-2-enyl anion in its 1st excited-state by (a) conrotatory, and (b) disrotatory, rotation.

A2.59 (a) Conrotatory: $\psi_1^2 \rightarrow \pi^2$, $\psi_2^1 \rightarrow \sigma^1$, $\psi_3^1 \rightarrow \sigma^{*1}$
$\therefore \psi_1^2\psi_2^1\psi_3^1$ (1st excited-state) $\rightarrow \sigma^1\pi^2\sigma^{*1}$ (highly excited-state).
(b) Disrotatory: $\psi_1^2 \rightarrow \sigma^2$, $\psi_2^1 \rightarrow \sigma^{*1}$, $\psi_3^1 \rightarrow \pi^1$
$\therefore \psi_1^2\psi_2^1\psi_3^1$ (1st excited-state) $\rightarrow \sigma^2\pi^1\sigma^{*1}$ (1st excited-state).

Q2.60 For the conversion of a prop-2-enyl anion into a cyclopropyl anion by a concerted, photochemical reaction (from the 1st excited-state of the prop-2-enyl anion), which mode of rotation would require the least activation energy from conservation of orbital symmetry considerations?

A2.60 Disrotatory rotation would require the least energy ($E_{\sigma^1\pi^2\sigma^{*1}} > E_{\sigma^2\pi^1\sigma^{*1}}$).

C2.4 Thermal interconversion of prop-2-enyl species and cyclopropyl species will occur *via* a disrotatory rotation for the cation, and *via* a conrotatory rotation for the anion. The photochemical interconversion (from the 1st excited-state) will occur *via* a conrotatory rotation for the cation, and *via* a disrotatory rotation for the anion. (The simple Frontier Orbital treatment gave the same predictions.)

Q2.61 Predict the geometry of a pent-3-en-2-yl cation formed by the thermal, electrocyclic opening of (a) *trans*-2,3-dimethylcyclopropyl cation, and (b) *cis*-2,3-dimethylcyclopropyl cation. Would there be more than one isomer formed in (b)?

A2.61 (a)

Me H + H Me $\xrightarrow[\text{DIS.}]{\Delta}$ H Me H + H Me (*A*)

(cont.)

(b)

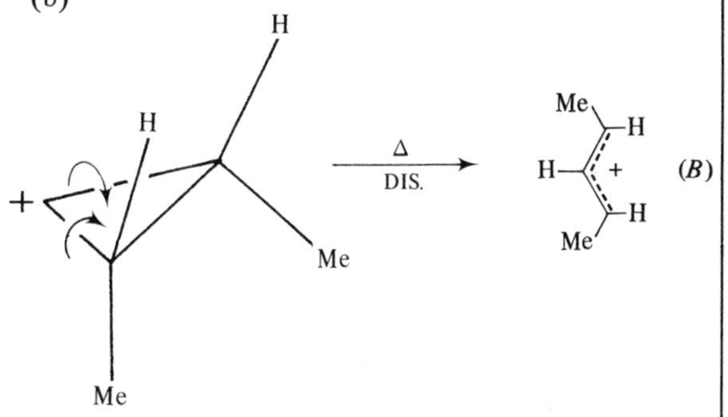

$\xrightarrow[\text{DIS.}]{\Delta}$ (B)

Steric interactions would make the alternative disrotatory rotation more difficult:

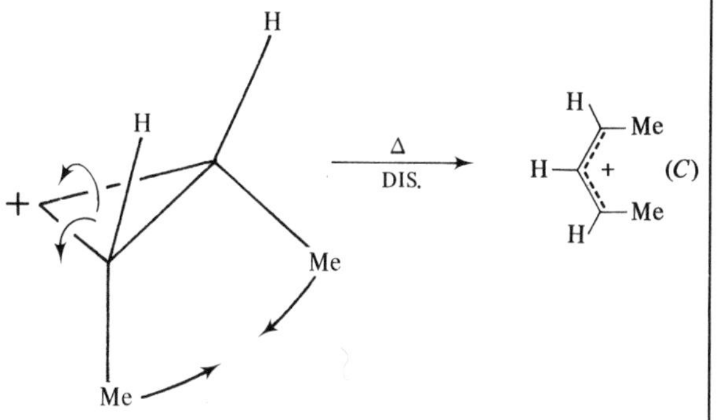

$\xrightarrow[\text{DIS.}]{\Delta}$ (C)

However, when the formation of the cyclopropyl cation by heterolytic cleavage of a σ-bond is synchronous with electrocyclic opening of the ring, the direction of disrotatory rotation is controlled by further stereo-electronic factors. The movement of the $C_{(2)}-C_{(3)}$ bond is predicted (by molecular orbital calculations)•, and found, to be in the opposite direction to that of the leaving group:

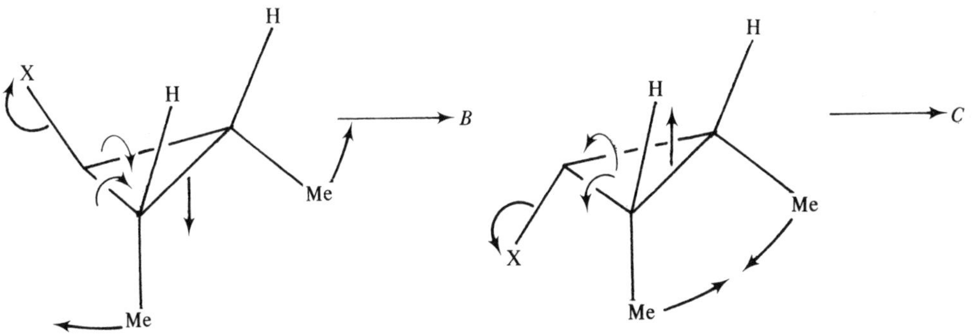

→ B

→ C

(P. von R. Schleyer, T. M. Su, M. Saunders and J. C. Rosenfeld, *J. Am. chem. Soc.*, 1969, **91**, 5174.)

• R. B. Woodward and R. Hoffmann, *J. Am. chem. Soc.*, 1965, **87**, 395.

A2.62

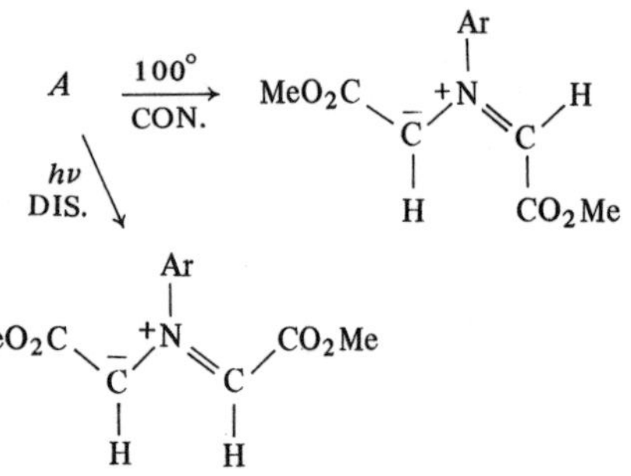

$A \xrightarrow[\text{CON.}]{100°}$ MeO_2C–$\bar{C}H$–$\overset{+}{N}(Ar)$=CH–CO_2Me

$A \xrightarrow[\text{DIS.}]{h\nu}$ MeO_2C–$\bar{C}H$–$\overset{+}{N}(Ar)$=CH–CO_2Me

(R. Huisgen, W. Scheer and H. Huber, *J. Am. chem. Soc.*, 1967, **89**, 1753.)

Q2.62 The electrocyclic opening of an aziridine to give an azomethine ylide:

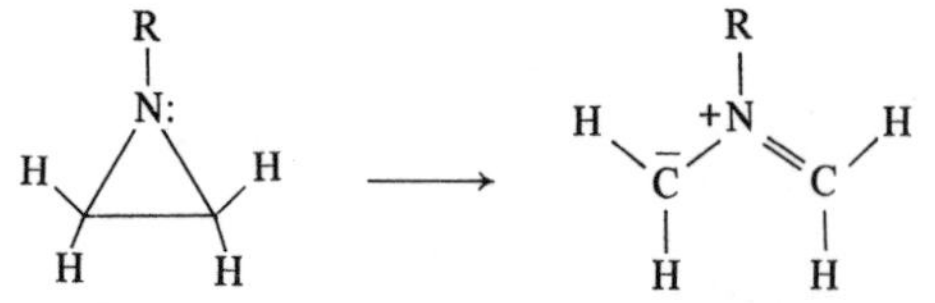

is iso-electronic with the cyclopropyl anion-prop-2-enyl anion transformation. Using this information, predict the geometry of the ylide formed (a) thermally, and (b) photochemically, from dimethyl *cis*-1-aryl aziridine-2,3-dicarboxylate (*A*).

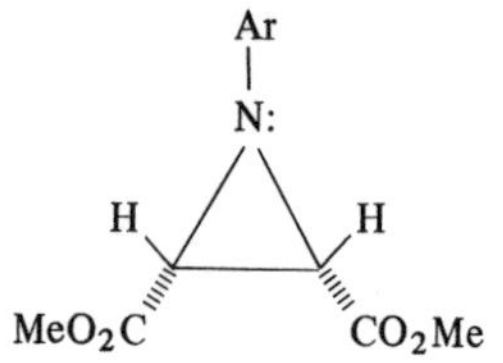

(*A*)

A2.63 $\sigma^2\pi^1$

Q2.63 What is the electronic configuration of the cyclopropyl radical in its ground-state?

A2.64 (a) Conrotatory: $\sigma^2 \rightarrow \psi_2^2$, $\pi^1 \rightarrow \psi_1^1$
$\therefore \sigma^2\pi^1$ (G.S.) $\rightarrow \psi_1^1\psi_2^2$ (an excited-state).
(b) Disrotatory: $\sigma^2 \rightarrow \psi_1^2$, $\pi^1 \rightarrow \psi_3^1$
$\therefore \sigma^2\pi^1$ (G.S.) $\rightarrow \psi_1^2\psi_3^1$ (an excited-state).

Q2.64 Using the orbital correlation diagrams, predict the electronic configuration of a prop-2-enyl radical derived from a cyclopropyl radical in its ground-state by (a) conrotatory, and (b) disrotatory, rotation.

S2.9 It is not obvious which of the two excited-states of the prop-2-enyl radical, $\psi_1^1\psi_2^2$ and $\psi_1^2\psi_3^1$, has the lowest energy. Therefore, prediction of the stereochemistry of the thermal interconversion of a prop-2-enyl radical and a cyclopropyl radical is somewhat ambiguous. The prediction for a photochemical interconversion is even more so.

S2.10 In the orbital correlation diagrams for the electrocyclic reactions: buta-1,3-diene-cyclobutene and prop-2-enyl-cyclopropyl, notice that when a reaction is symmetry forbidden it is the correlation of the *highest occupied molecular orbital* of the reactant with a high-energy orbital of the intended product which causes the large increase in activation energy. This gives us a rationale for the success of the Frontier Orbital approach.

3 Cyclo-addition reactions

Introduction

In this section we shall use an orbital correlation approach to test for conservation of orbital symmetry in cyclo-addition reactions.•

We can define a cyclo-addition reaction as the formation of σ-bonds between the termini of two (or more) π-systems:

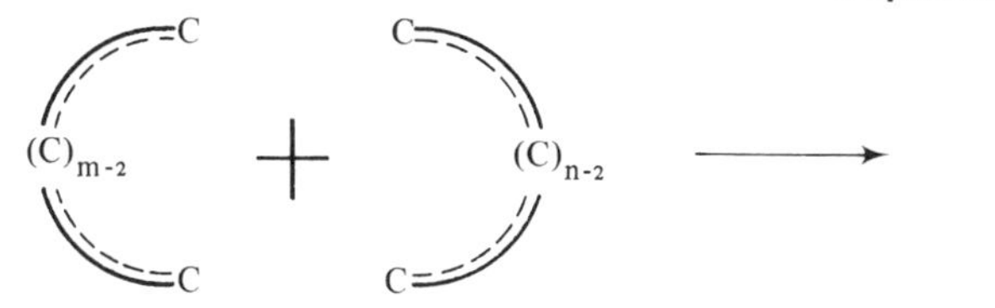

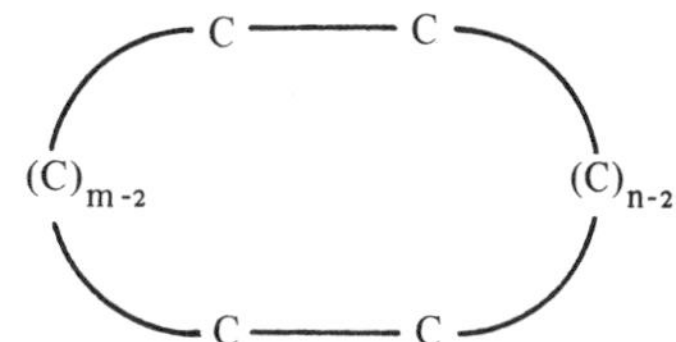

One of the most well known cyclo-addition reactions is the Diels–Alder reaction, which in essence involves the thermal addition of a conjugated diene to a double bond (a 4π-electron + 2π-electron cyclo-addition, or 4 + 2 cyclo-addition):

We shall analyse this reaction first.

S3.1 The most energetically favourable arrangement (giving most effective overlap of the interacting orbitals) for cyclo-addition is a face-to-face approach of the two reactants, as shown. This is known as a suprafacial-suprafacial cyclo-addition since the two new σ-bonds to each reactant are formed on one face only of each reactant. In this geometry, a plane of symmetry (m) is preserved throughout the reaction. This lies perpendicular to the planes of both reacting species and bisects them.

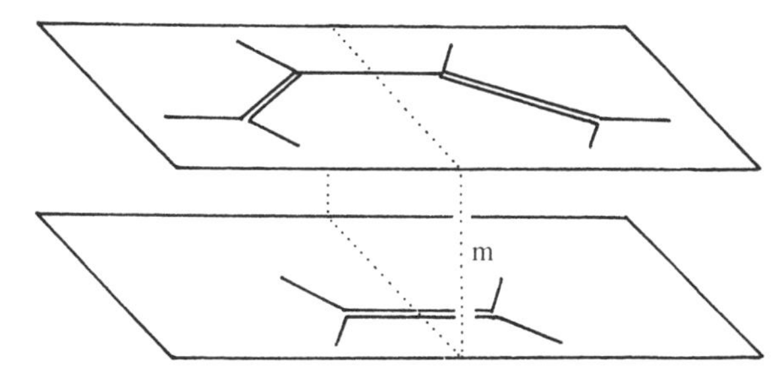

The molecular orbitals which are pertinent to the analysis are: ψ_1, ψ_2, ψ_3, and ψ_4 of the diene, π and π^* of the olefin, and σ_1, σ_2, σ_3, σ_4, π and π^* of the cyclohexene which is formed. The σ-orbitals of the cyclohexene which are relevant are those which are formed during the reaction. However, the localised orbitals:

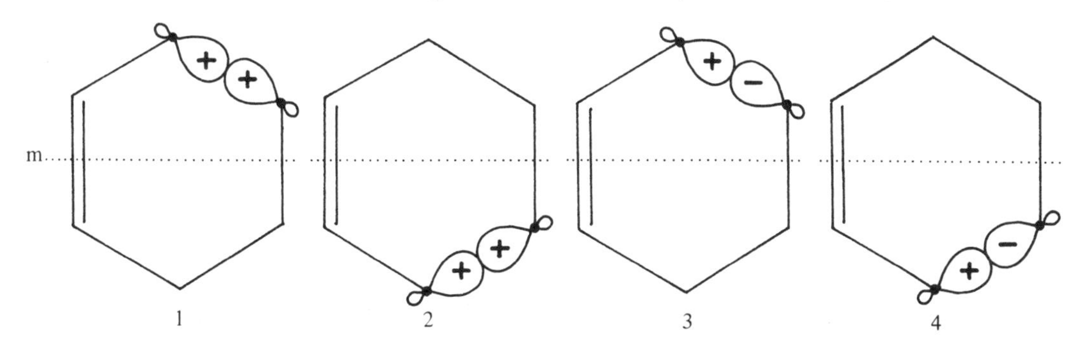

• By the principle of microscopic reversibility, the reverse reactions will be subject to the same orbital symmetry restraints, and will follow the same stereochemical course.

are of no use since they are neither symmetric nor antisymmetric with respect to the plane of symmetry, m. Instead, we must use linear combinations of these orbitals:

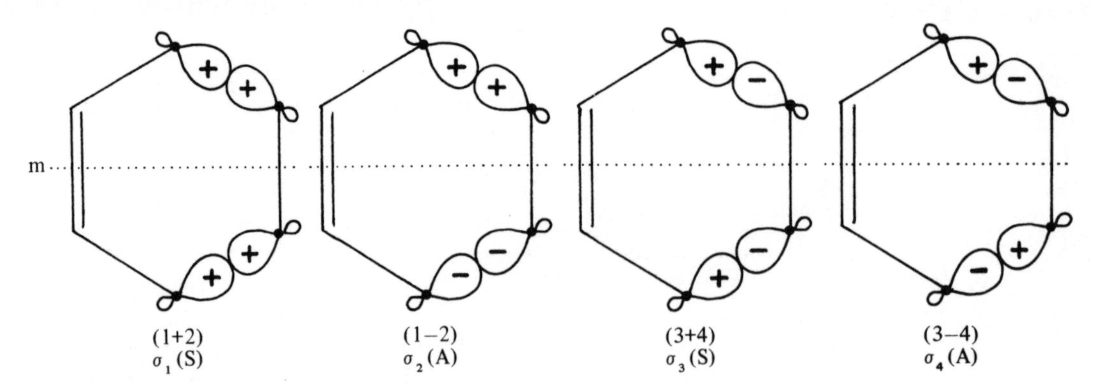

These 'new' orbitals, in fact, are probably a better representation of the real system than the localised σ-bonds.• The energy of these orbitals increases from σ_1 to σ_4, σ_1 and σ_2 being bonding orbitals, and σ_3 and σ_4 being antibonding orbitals. Their symmetry classification with respect to m is indicated.

Q3.1 Classify the symmetry of ψ_1, ψ_2, ψ_3, and ψ_4 of the diene, and π and π^* of the olefin and cyclohexene, with respect to the plane of symmetry m.

A3.1 Diene: ψ_1 (S), ψ_2 (A), ψ_3 (S), ψ_4 (A)
Olefin: π (S), π^* (A)
Cyclohexene: π (S), π^* (A)

Q3.2 Collect together the symmetric and the antisymmetric orbitals in the reactants, and in the product.

• The molecular orbitals of buta-1,3-diene can be constructed from the localised π and π^* orbitals of the two adjacent double bonds in the same way:

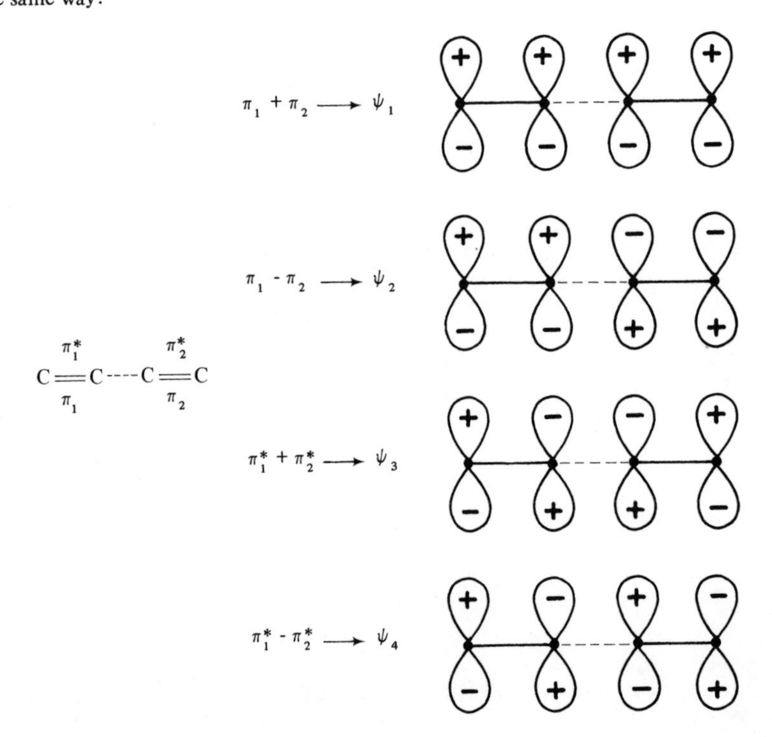

A3.2

	Symmetric orbitals	*Antisymmetric orbitals*
Reactants:	ψ_1, ψ_3, π	ψ_2, ψ_4, π^*
Product:	σ_1, σ_3, π	$\sigma_2, \sigma_4, \pi^*$

S3.2 During the course of the reaction, the symmetric orbitals of the reactants interact and mix together, giving the symmetric orbitals of the product. Similarly, the antisymmetric orbitals of the reactants mix to give the antisymmetric orbitals of the product. A pictorial representation of the mixing of the orbitals is not so readily obtained in this case, cf. the buta-1,3-diene-cyclobutene electrocyclic reaction (Chapter 2, pp. 22–27), as three orbitals are involved, but the same general principles apply, viz. orbitals of lower energy mix in those of higher energy in a bonding manner, while orbitals of higher energy mix in those of lower energy in an antibonding manner:

Symmetric orbitals

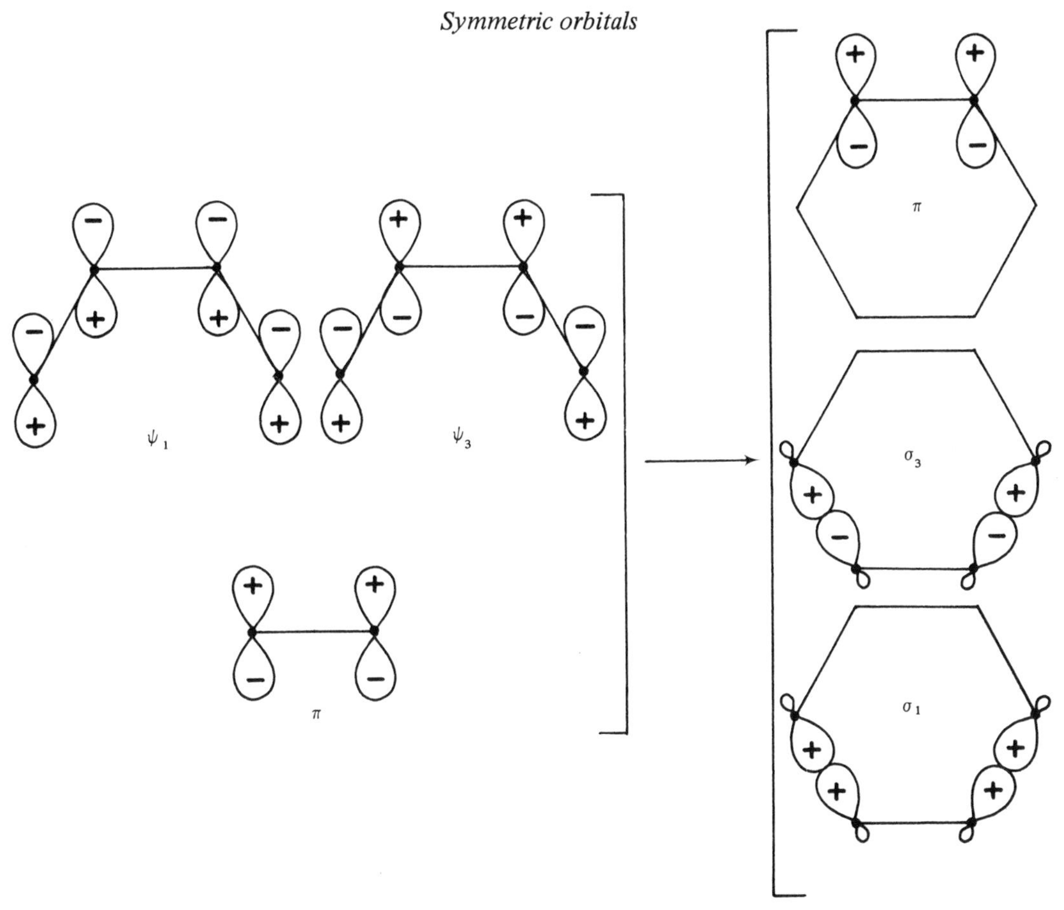

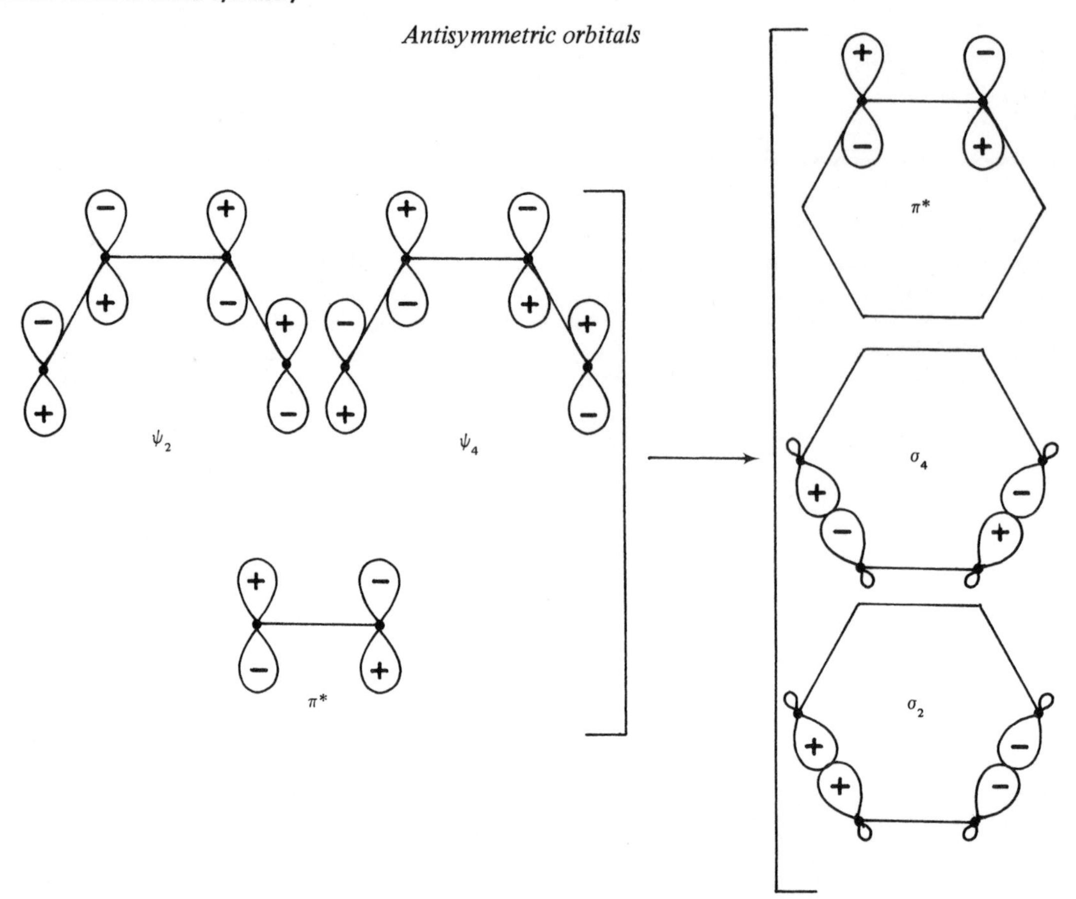

Q3.3 Construct an orbital correlation diagram for a diene + olefin cyclo-addition, by (i) arranging the orbitals of the reactants on the left, in order of increasing energy (orbital of lowest energy at the bottom), (ii) arranging the orbitals of the product on the right, in order of increasing energy, (iii) indicating the symmetry classification of each orbital with respect to the plane of symmetry, and (iv) connecting orbitals with the same symmetry classification in the reactants and the product, bearing in mind that lines joining two pairs of orbitals of the same symmetry must not cross.

A3.3

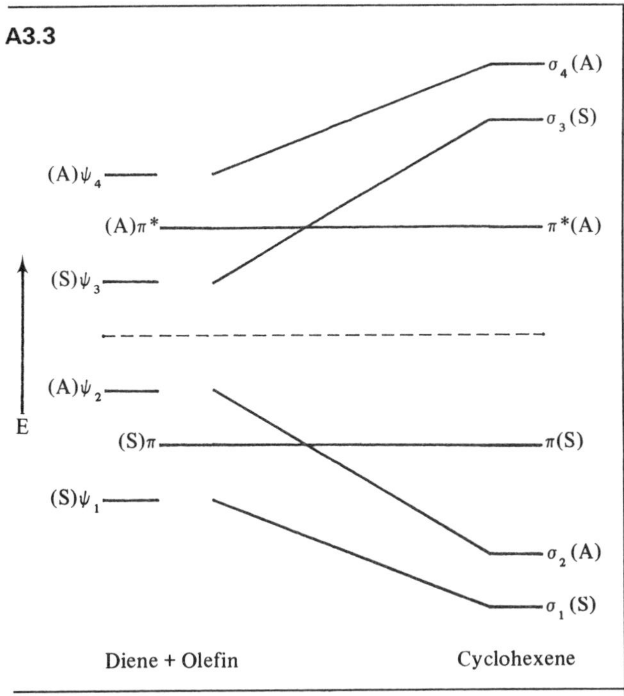

Q3.4 What is the electronic configuration of the reactants in their ground-state?

A3.4 Diene: $\psi_1^2\psi_2^2$ Olefin: π^2

Q3.5 Starting with the reactants in their ground-state electronic configuration, what electronic configuration of the cyclohexene would be obtained? Is this the ground-state of cyclohexene?

A3.5 $\psi_1^2 \to \sigma_1^2$, $\psi_2^2 \to \sigma_2^2$, $\pi^2 \to \pi^2$

$\therefore \psi_1^2\psi_2^2\pi^2 \to \sigma_1^2\sigma_2^2\pi^2$ (the ground-state electronic configuration of cyclohexene). The reaction is '*allowed*'.

Q3.6 If cyclo-addition occurred between the diene in its 1st excited-state and the olefin in its ground-state, what electronic configuration of the cyclohexene would be obtained? Is this reaction likely to occur?

A3.6 Diene: $\psi_1^2\psi_2^1\psi_3^1$ Olefin: π^2

$\psi_1^2 \to \sigma_1^2$, $\psi_2^1 \to \sigma_2^1$, $\psi_3^1 \to \sigma_3^1$, $\pi^2 \to \pi^2$

$\therefore \psi_1^2\psi_2^1\psi_3^1\pi^2 \to \sigma_1^2\sigma_2^1\pi^2\sigma_3^1$ (a highly excited-state of cyclohexene).

A considerable amount of energy would be required to attain such a state even after the initial electronic excitation, therefore the reaction is unlikely to occur.

Q3.7 If cyclo-addition occurred between the diene in its ground-state and the olefin in its 1st excited-state, what electronic configuration of the cyclohexene would be obtained? Is this reaction likely to occur?

A3.7 Diene: $\psi_1^2\psi_2^2$ Olefin: $\pi^1\pi^{*1}$

$\psi_1^2 \to \sigma_1^2$, $\psi_2^2 \to \sigma_2^2$, $\pi^1 \to \pi^1$, $\pi^{*1} \to \pi^{*1}$

$\therefore \psi_1^2\psi_2^2\pi^1\pi^{*1} \to \sigma_1^2\sigma_2^2\pi^1\pi^{*1}$ (1st excited-state of cyclohexene).

From conservation of orbital symmetry considerations, little energy would be required to achieve this reaction *once the initial electronic excitation had been achieved.* A more detailed analysis (Chapter 5, page 74) reveals, however, that this correlation is in fact symmetry *forbidden,* i.e. $\psi_1^2\psi_2^2\pi^1\pi^{*1} \nrightarrow \sigma_1^2\sigma_2^2\pi^1\pi^{*1}$.

Q3.8 If the thermal cyclo-addition of *trans,trans*-hexa-2,4-diene and *trans*-but-2-ene occurs in a concerted manner through a suprafacial-suprafacial arrangement, what would be the expected stereochemistry of the cyclohexene derivative formed?

A3.8

C3.1 The thermal Diels-Alder reaction satisfies the principle of conservation of orbital symmetry.

Q3.9 Which are the relevant molecular orbitals for the analysis of a suprafacial-suprafacial cyclo-addition of a prop-2-enyl species and an olefin, giving a cyclopentyl species? (*N.B.* The number of orbitals in the product must equal those in the reactants.)

A3.9

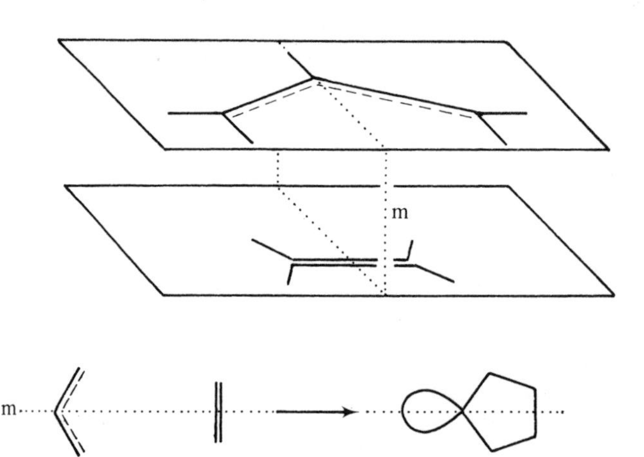

Prop-2-enyl orbitals: ψ_1, ψ_2 and ψ_3
Olefin orbitals: π and π^*
Cyclopentyl orbitals: $\sigma_1, \sigma_2, \sigma_3, \sigma_4$ (cf. S3.1) and n (non-bonding orbital on the carbon which was $C_{(2)}$ of the prop-2-enyl species; this may be an sp^3 or a p atomic orbital).

Q3.10 Classify the symmetry of these orbitals with respect to the plane of symmetry, *m*.

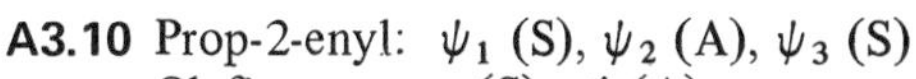

A3.10 Prop-2-enyl: ψ_1 (S), ψ_2 (A), ψ_3 (S)
Olefin: π (S), π^* (A)
Cyclopentyl: σ_1 (S), σ_2 (A), σ_3 (S), σ_4 (A), n(S)

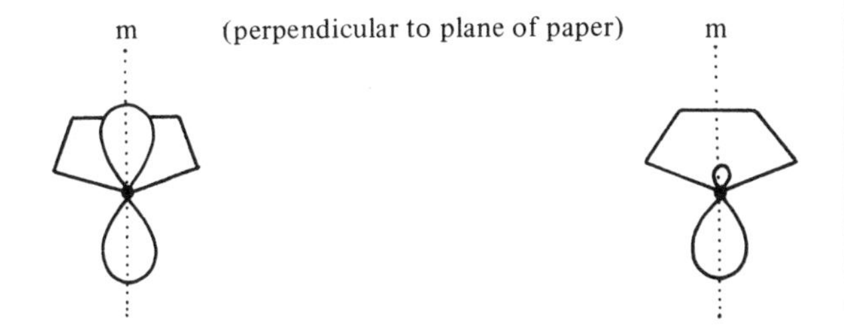

Non-bonding orbital (n) bisected by plane of symmetry ∴ symmetric.

Q3.11 Construct an orbital correlation diagram for the suprafacial-suprafacial cyclo-addition of a prop-2-enyl species and an olefin, to give a cyclopentyl species.

A3.11

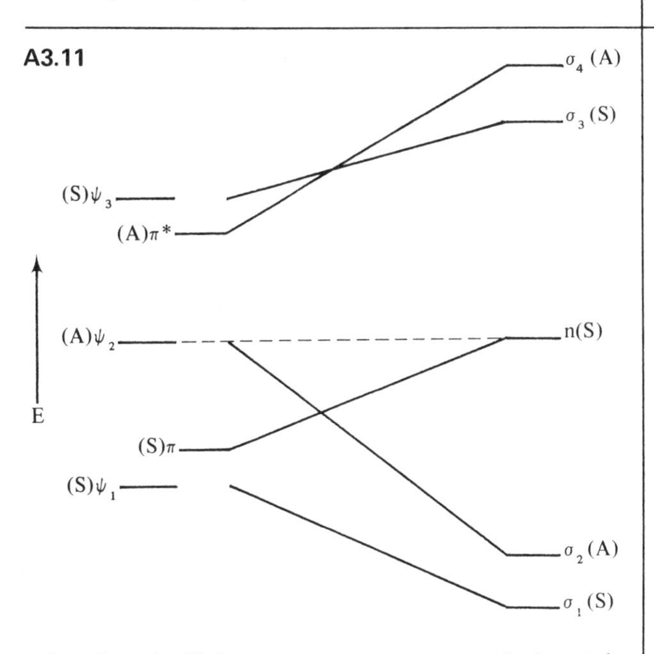

Q3.12 For the suprafacial-suprafacial cyclo-addition of a prop-2-enyl *cation* and an olefin, both in their ground-state electronic configurations, what would be the electronic configuration of the cyclopentyl *cation* formed? Is this reaction likely to occur?

A3.12 Prop-2-enyl cation: ψ_1^2 Olefin: π^2
$\psi_1^2 \rightarrow \sigma_1^2,\ \pi^2 \rightarrow n^2$
$\therefore\ \psi_1^2\pi^2 \rightarrow \sigma_1^2 n^2$ (an excited-state of the cyclopentyl cation).

The ground-state of the product would be $\sigma_1^2\sigma_2^2$. The reaction, a thermal 2 + 2 cyclo-addition, is unlikely to occur (in a concerted manner).

Q3.13 For the suprafacial-suprafacial cyclo-addition of a prop-2-enyl *anion* and an olefin, both in their ground-state electronic configurations, what would be the electronic configuration of the cyclopentyl *anion* formed? Is this reaction likely to occur?

A3.13 Prop-2-enyl anion: $\psi_1^2\psi_2^2$ Olefin: π^2
$\psi_1^2 \rightarrow \sigma_1^2,\ \psi_2^2 \rightarrow \sigma_2^2,\ \pi^2 \rightarrow n^2$
$\therefore\ \psi_1^2\psi_2^2\pi^2 \rightarrow \sigma_1^2\sigma_2^2 n^2$ (G.S. of the cyclopentyl anion).

From conservation of orbital symmetry considerations, little energy would be required to achieve this reaction, which is a thermal 4 + 2 cyclo-addition.

Q3.14 Referring back to Q2.62, p. 36, what would be the stereochemistry of the products formed on trapping the intermediate azomethine ylides by cyclo-addition with dimethyl acetylenedicarboxylate?

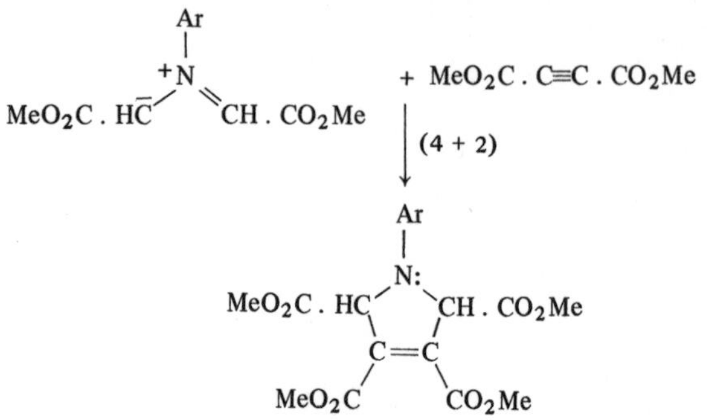

Ar
+N
$MeO_2C.H\bar{C}$ $CH.CO_2Me$ + $MeO_2C.C{\equiv}C.CO_2Me$
(4 + 2)
Ar
N:
$MeO_2C.HC$ $CH.CO_2Me$
C=C
MeO_2C CO_2Me

[*N.B.* Azomethine ylides and 2,5-dihydropyrrole derivatives are isoelectronic with the prop-2-enyl anion and the cyclopent-3-enyl anion respectively. Only two π-electrons of the acetylene are involved in the cyclo-addition, therefore the other two π-electrons may be ignored, both in the reactant and the product (the $C_{(3)}C_{(4)}$ double bond in the latter).]

A3.14

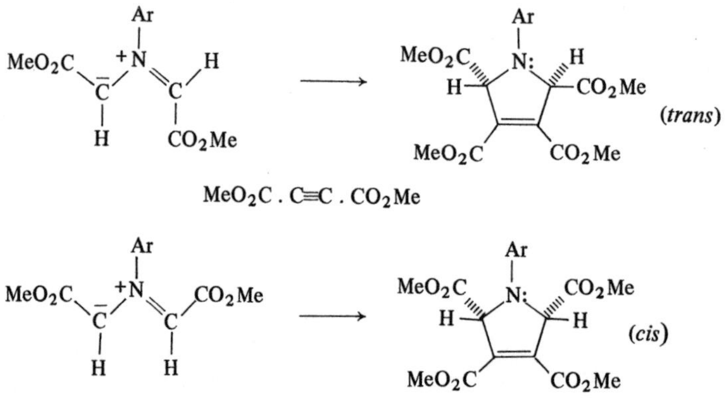

Ar, +N, MeO_2C, H, $\bar{C}$, C, H, CO_2Me → Ar, N:, MeO_2C, H, H, CO_2Me, MeO_2C, CO_2Me (*trans*)

$MeO_2C.C{\equiv}C.CO_2Me$

Ar, +N, MeO_2C, CO_2Me, $\bar{C}$, C, H, H → Ar, N:, MeO_2C, CO_2Me, H, H, MeO_2C, CO_2Me (*cis*)

The stereochemistry of the cyclo-adducts gives directly the geometry of the intermediate azomethine ylides. The latter are not isolated.

(R. Huisgen, W. Scheer and H. Huber, *J. Am. chem. Soc.*, 1967, **89**, 1753.)

Q3.15 Which are the relevant molecular orbitals for the analysis of a suprafacial-suprafacial cyclo-addition of two olefins giving a cyclobutane?

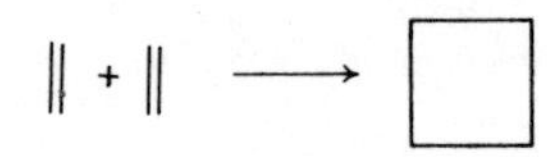

|| + || → □

A3.15

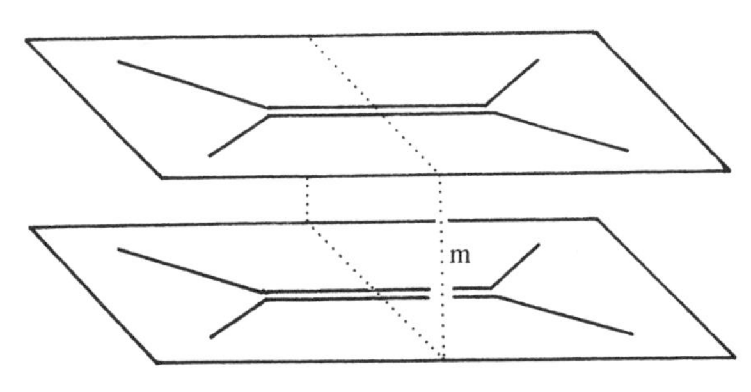

Olefin orbitals: 2π and $2\pi^*$
Cyclobutane orbitals: $\sigma_1, \sigma_2, \sigma_3, \sigma_4$ (cf. S3.1).

Q3.16 Classify the symmetry of these orbitals with respect to the plane of symmetry, *m*.

A3.16 Olefins: π (S), π (S), π^* (A), π^* (A)
Cyclobutane: σ_1 (S), σ_2 (A), σ_3 (S), σ_4 (A)

Q3.17 Construct an orbital correlation diagram for the suprafacial-suprafacial cyclo-addition of two olefins to give a cyclobutane.

A3.17

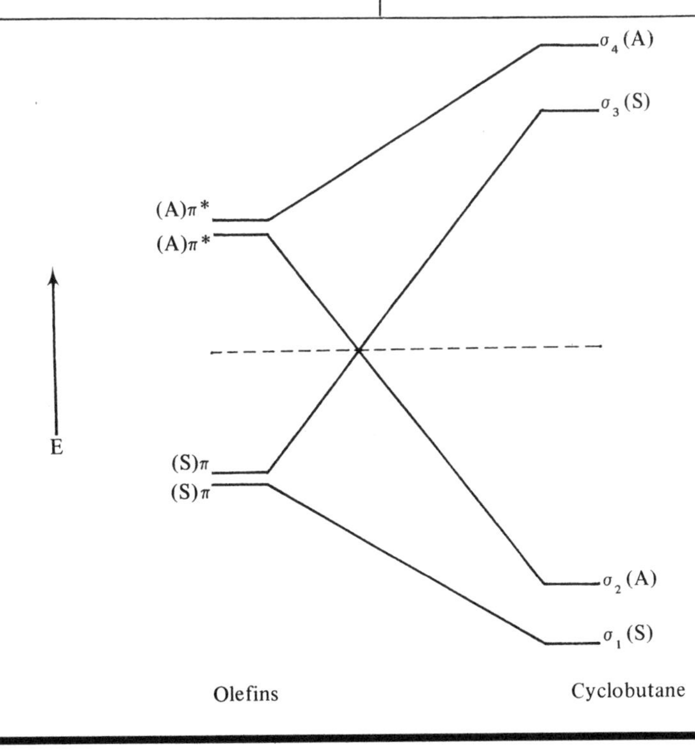

S3.3 Mixing of the symmetric orbitals of the olefins, π and π, results in the formation of the symmetric orbitals of the cyclobutane, σ_1 and σ_3. Similarly, mixing of the antisymmetric orbitals, π^* and π^*, leads to the antisymmetric orbitals of the cyclobutane, σ_2 and σ_4. For this system, this can be conveniently represented pictorially:

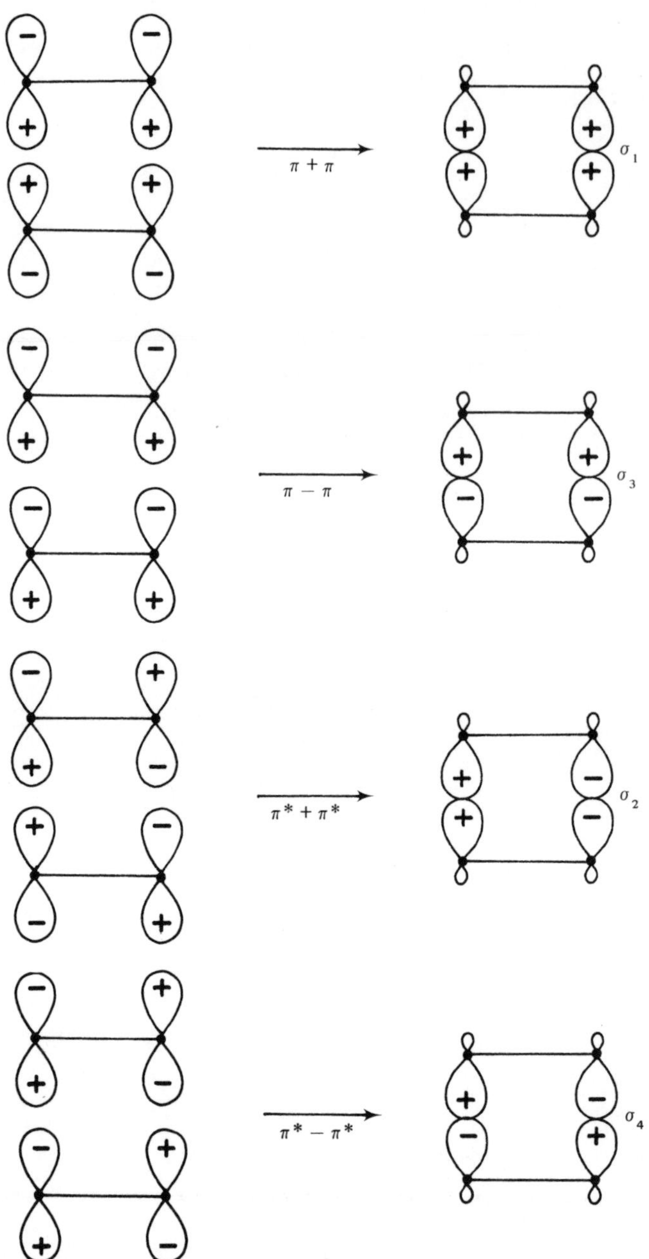

A rigorous analysis would also consider orbital symmetry with respect to a second plane of symmetry for this particular reaction, which lies parallel to, and midway between, the planes of the two reacting olefins. However, the orbital correlation obtained by this approach is the same as in our simplified analysis.

Q3.18 For the suprafacial-suprafacial cyclo-addition of two olefins, both in their ground-state electronic configuration, what would be the electronic configuration of the cyclobutane formed? Is this reaction likely to occur?

A3.18 $\pi^2 \xrightarrow{(\pi + \pi)} \sigma_1^2,\ \pi^2 \xrightarrow{(\pi - \pi)} \sigma_3^2$

$\therefore \pi^2\pi^2 \rightarrow \sigma_1^2\sigma_3^2$ (an excited-state of cyclobutane). The reaction, a thermal 2 + 2 cyclo-addition, is unlikely to occur.

Q3.19 For the suprafacial-suprafacial cyclo-addition of two olefins, one of which is in its 1st excited-state, the other being in its ground-state, what would be the electronic configuration of the cyclobutane formed? Is this reaction likely to occur?

A3.19 Olefin 1: $\pi^1\pi^{*1}$ Olefin 2: π^2

$\pi^2 \xrightarrow{(\pi + \pi)} \sigma_1^2,\ \pi^1 \xrightarrow{(\pi - \pi)} \sigma_3^1,\ \pi^{*1} \xrightarrow{(\pi^* + \pi^*)} \sigma_2^1$

$\therefore \pi^2\pi^1\pi^{*1} \rightarrow \sigma_1^2\sigma_2^1\sigma_3^1$ (the 1st excited-state of cyclobutane).

From conservation of orbital symmetry considerations, little energy would be required to achieve this reaction, a photochemical 2 + 2 cyclo-addition, once the initial electronic excitation had been achieved.

Q3.20 Which stereoisomers of 1,2,3,4-tetramethylcyclobutane would you expect to be formed on photolysis of (a) *cis*-but-2-ene, (b) *trans*-but-2-ene, and (c) a mixture of *cis*- and *trans*-but-2-ene? Assume that no photo-isomerisation of the reactants occurs.

A3.20 Suprafacial-suprafacial, photochemical 2 + 2 cyclo-addition

(a)

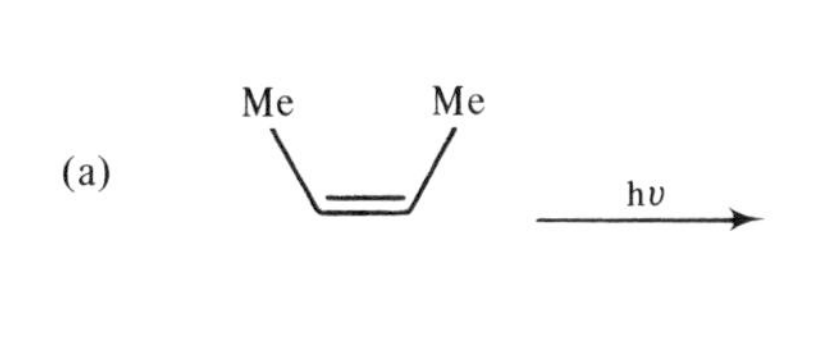

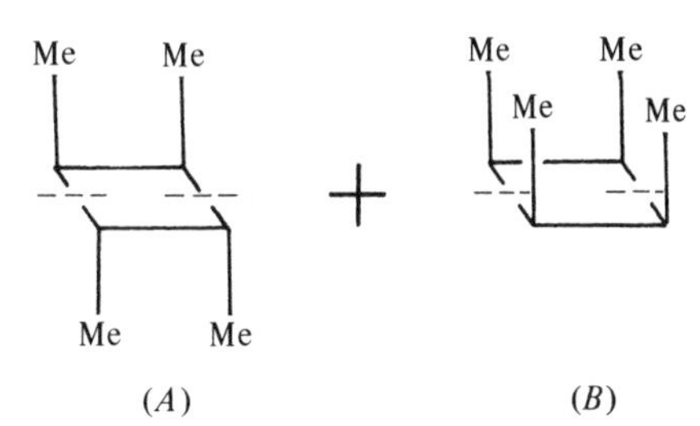

(b) Me Me hυ

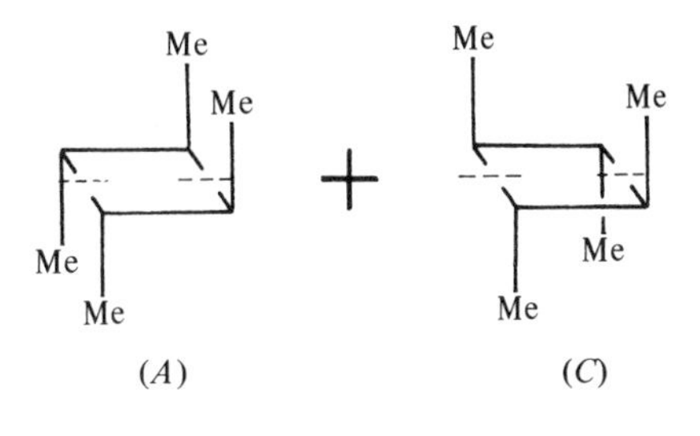

(c) Me Me + Me Me hυ

A + *B* from *cis*-but-2-ene

A + *C* from *trans*-but-2-ene

+

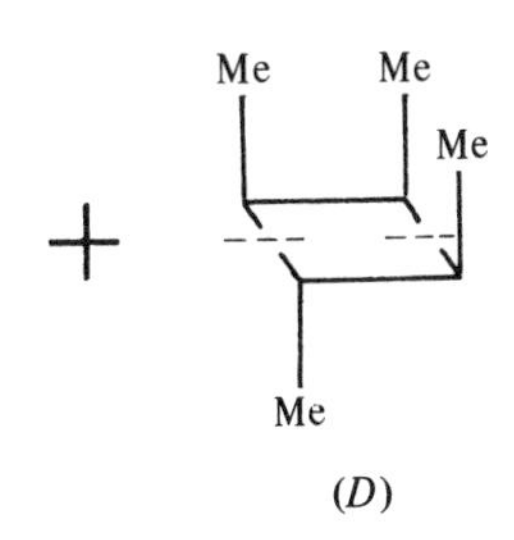

(H. Yamazaki and R. J. Cvetanovic, *J. Am. chem. Soc.*, 1969, 91, 520.)

C3.2 *Predictions for suprafacial-suprafacial cyclo-additions*

Reactants	Thermal (both reactants in their ground-states)	Photochemical (one reactant in its 1st excited-state; the other in its ground-state)
C=C + C=C (2 + 2)	Forbidden	Allowed
C=C + C=C–C$^+$ (2 + 2)	Forbidden	Allowed
C=C + C=C–C$^-$ (2 + 4)	Allowed	Forbidden
C=C + C=C–C=C (2 + 4)	Allowed	Forbidden
C=C–C$^+$ + C=C–C=C (2 + 4)	Allowed	Forbidden
C=C–C$^-$ + C=C–C=C (4 + 4)	Forbidden	Allowed
Total no. of π-electrons		
$4n$	Forbidden	Allowed
$4n + 2$	Allowed	Forbidden
(n = 1, 2, 3, ...)		

S3.4 Although the *suprafacial-suprafacial* arrangement for a cyclo-addition reaction appears to be preferred if reaction in this mode can occur with conservation of orbital symmetry, it is not the only possibility. Alternative modes are (i) *suprafacial-antarafacial,* in which the new σ-bonds are formed to one face only of one reactant (suprafacially), and to both faces of the other reactant (antarafacially), and (ii) *antarafacial-antarafacial,* in which the new σ-bonds are formed to both faces of both reactants (both antarafacially). These modes of cyclo-addition are illustrated for the formation of a cyclobutane by the cyclo-addition of two olefins, a reaction which is symmetry forbidden in the suprafacial-suprafacial mode.

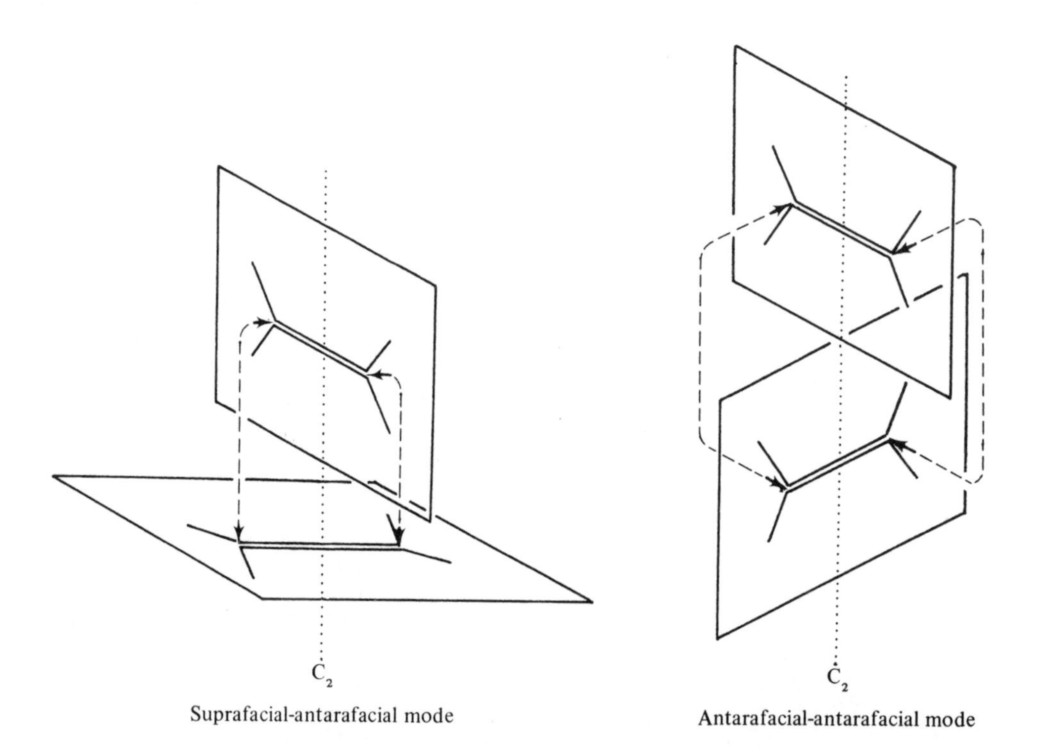

Suprafacial-antarafacial mode

Antarafacial-antarafacial mode

Dashed arrows represent orbital interactions leading to σ-bonding. Both of these arrangements have a 2-fold axis of symmetry as indicated.

The effectiveness with which the appropriate orbitals can overlap in these arrangements is however much less than in the suprafacial–suprafacial arrangement, and for this reason reaction through these pathways is not often observed, even when it is symmetry allowed.

Q3.21 For a *suprafacial–antarafacial* cyclo-addition of two olefins, forming a cyclobutane, which are the relevant orbitals for the analysis of conservation of orbital symmetry in this reaction?

A3.21 π and π^* of both olefins
σ_1, σ_2, σ_3, and σ_4 of the cyclobutane (cf. S3.1).

Q3.22 With the help of the diagrams given below, classify these orbitals according to their symmetry with respect to the 2-fold axis of symmetry present in a suprafacial–antarafacial cyclo-addition.

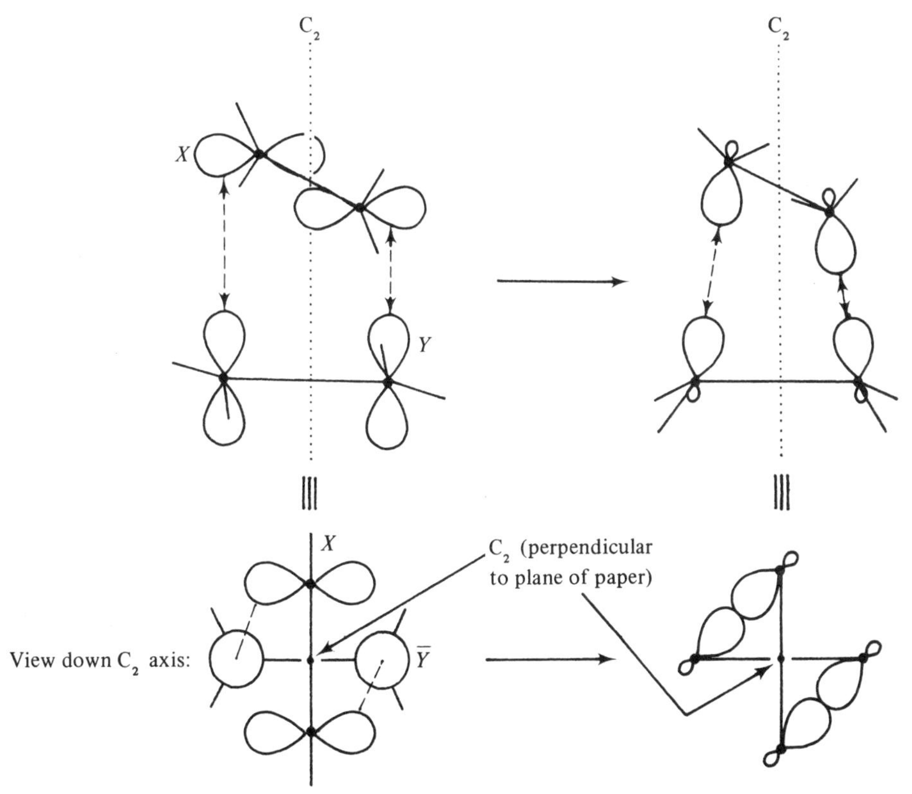

A3.22 Olefin X (antarafacial): π (A), π^* (S)
Olefin Y (suprafacial): π (S), π^* (A)
Cyclobutane: σ_1 (S), σ_2 (A), σ_3 (S), σ_4 (A)

Q3.23 Using the diagrams given in the previous question draw the interactions of (a) the symmetric orbitals of the reactants giving the symmetric orbitals of the product, and (b) the antisymmetric orbitals of the reactants giving the antisymmetric orbitals of the product.

A3.23 (a) Symmetric orbitals

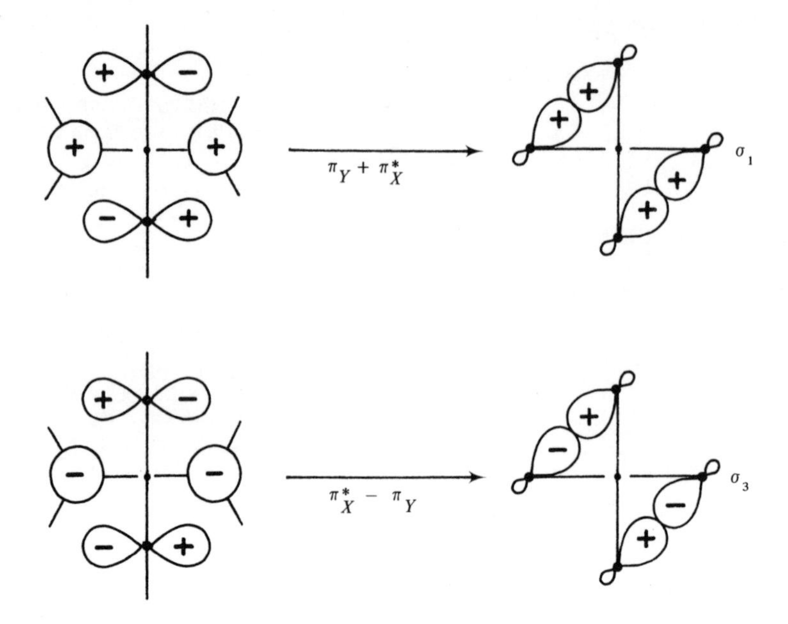

(b) Antisymmetric orbitals

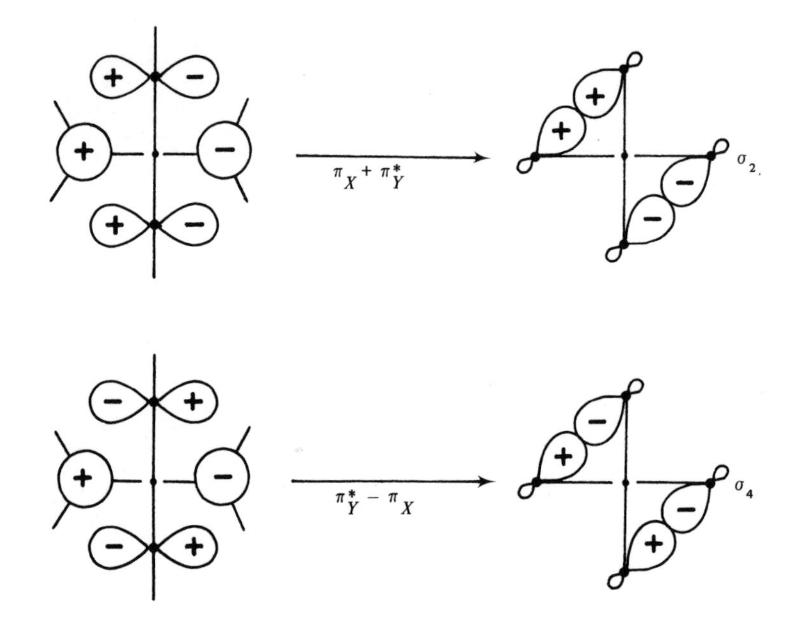

Q3.24 Construct an orbital correlation diagram for the suprafacial-antarafacial cyclo-addition of two olefins to give a cyclobutane.

A3.24

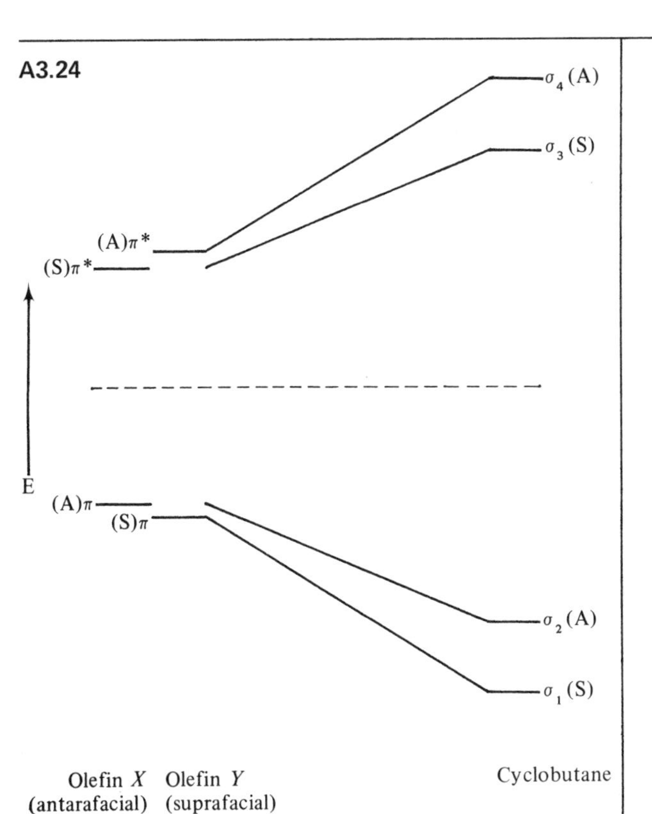

Q3.25 What would be the electronic configuration of the cyclobutane formed by a suprafacial-antarafacial cyclo-addition of two olefins, both in their ground-state electronic configuration?

A3.25 π^2 (suprafacial) $\rightarrow \sigma_1^2$
π^2 (antarafacial) $\rightarrow \sigma_2^2$
$\therefore \pi^2_{(s)}\pi^2_{(a)} \rightarrow \sigma_1^2\sigma_2^2$ (the ground-state electronic configuration of cyclobutane). The reaction is 'allowed'.

Q3.26 *If* a concerted, thermal, suprafacial–antarafacial cyclo-addition of two molecules of *cis*-but-2-ene could be realised, which stereoisomer of 1, 2, 3, 4-tetramethylcyclobutane would be formed?

A3.26

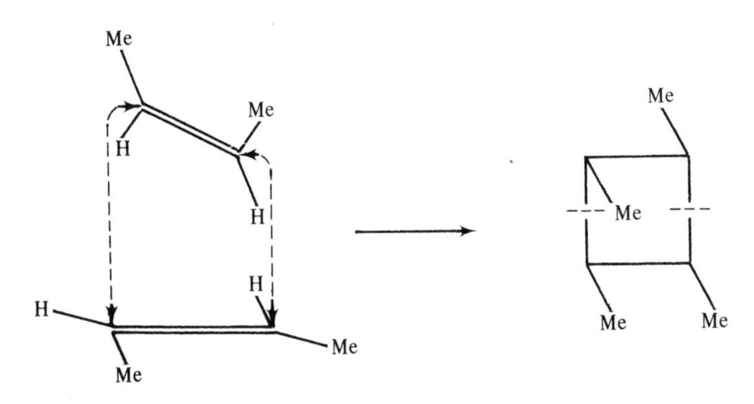

Q3.27 If the thermal cleavage of *cis*-bicyclo[4,2,0]octane to give cyclohexene and ethylene occurs by a concerted process, a suprafacial-antarafacial pathway must be followed. For the deuterium labelled system:

H D H H 450° H H + CHD=CHD H D

what would be the expected geometry of the ethylene-d_2?

A3.27 Assuming the cyclohexene to be the suprafacial component (the geometry at the ring-junction is preserved), then the CHD=CHD will be the antarafacial component. The relative geometry at $C_{(7)}$ and $C_{(8)}$ in the reactant will therefore become inverted during the reaction, and the ethylene-d_2 should have a *trans*-configuration.

(J. E. Baldwin and P. W. Ford, *J. Am. chem. Soc.*, 1969, 91, 7192.)

C3.3 The thermal cyclo-addition of two olefins to form a cyclobutane by a suprafacial-antarafacial mode is symmetry allowed.

S3.5 The occurrence of suprafacial-antarafacial cyclo-additions becomes more probable (if orbital symmetry is conserved) with higher polyene systems since these can accommodate more readily the distortion from planarity which is necessary to get effective overlap of the interacting orbitals in this mode of reaction.

Predictions for suprafacial-antarafacial cyclo-additions

Reactants	Thermal (both reactants in their ground-states)	Photochemical (one reactant in its 1st excited-state; the other in its ground-state)
C=C + C=C (2 + 2)	Allowed	Forbidden
C=C + C=C–C$^+$ (2 + 2)	Allowed	Forbidden
C=C + C=C–C$^-$ (2 + 4)	Forbidden	Allowed
C=C + C=C–C=C (2 + 4)	Forbidden	Allowed
C=C–C$^+$ + C=C–C=C (2 + 4)	Forbidden	Allowed
C=C–C$^-$ + C=C–C=C (4 + 4)	Allowed	Forbidden
Total no. of π-electrons		
$4n$	Allowed	Forbidden
$4n + 2$	Forbidden	Allowed
($n = 1, 2, 3, \ldots$)		

(Either of the reactants may be the antarafacial component).

Cf. Corresponding Table for suprafacial–suprafacial cyclo-additions (C3.2, page 48); reactants for which the suprafacial–suprafacial mode of cyclo-addition is symmetry forbidden are allowed to react by the suprafacial-antarafacial mode, and *vice versa*.

Q3.28 For an *antarafacial-antarafacial* cyclo-addition of two olefins, classify the symmetry of the orbitals π and π^* of both olefins, and σ_1, σ_2, σ_3, and σ_4 of the cyclobutane formed, with respect to the 2-fold axis of symmetry present in this arrangement.

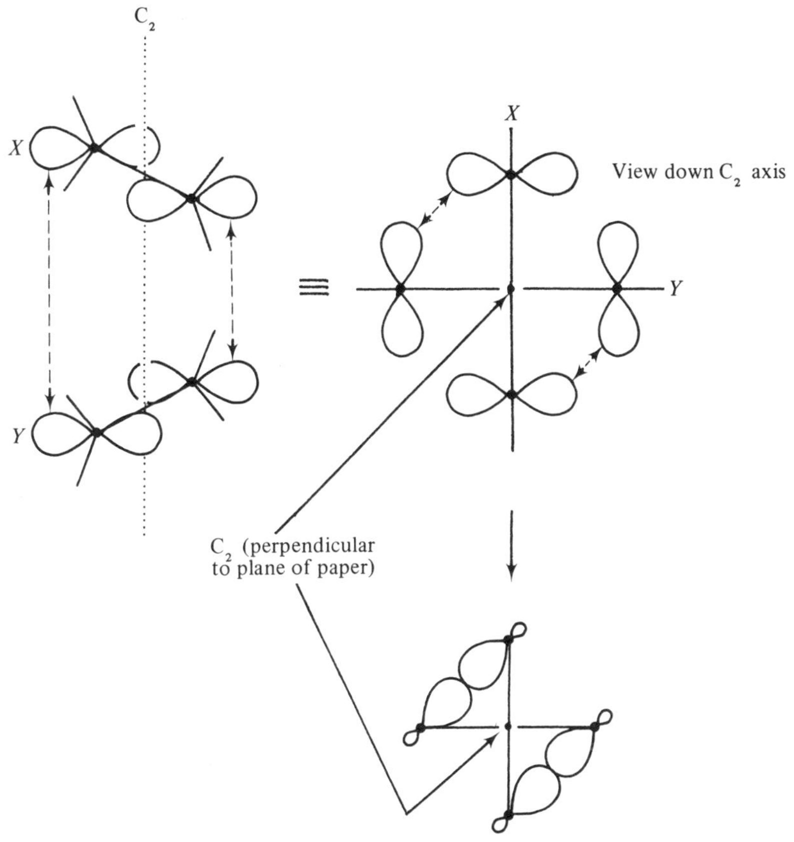

A3.28 Olefin X (antarafacial): π (A), π^* (S)
Olefin Y (antarafacial): π (A), π^* (S)
Cyclobutane: σ_1 (S), σ_2 (A), σ_3 (S), σ_4 (A)

Q3.29 Construct an orbital correlation diagram for the antarafacial-antarafacial cyclo-addition of two olefins to form a cyclobutane.

A3.29

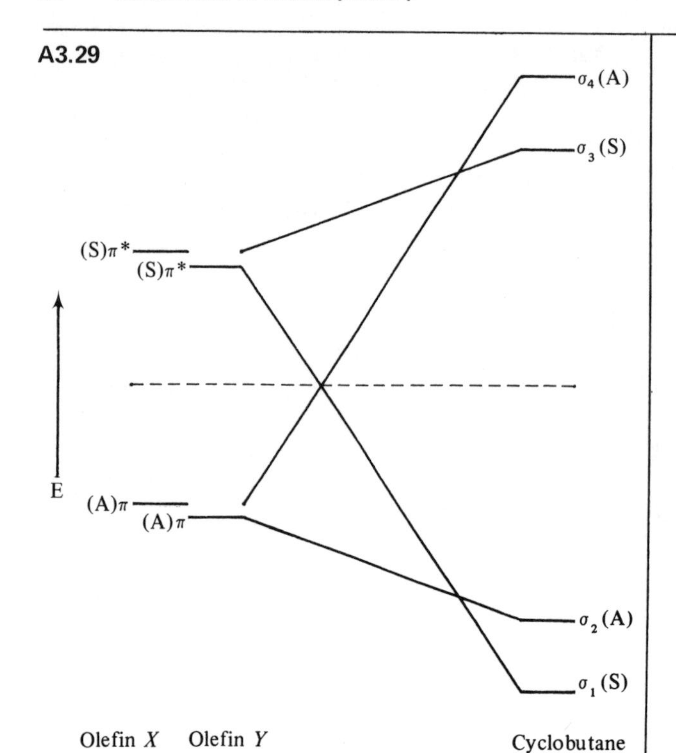

Q3.30 What would be the electronic configuration of the cyclobutane formed by an antarafacial-antarafacial cyclo-addition of two olefins, both in their ground-state electronic configurations?

A3.30 $\pi^2 \to \sigma_2^2$, $\pi^2 \to \sigma_4^2$

$\therefore \pi_{(a)}^2\pi_{(a)}^2 \to \sigma_2^2\sigma_4^2$ (an excited-state of cyclobutane).

Reaction by an antarafacial–antarafacial mode to give the ground-state of cyclobutane cannot therefore occur with conservation of orbital symmetry.

S3.6 The predictions for antarafacial–antarafacial cyclo-additions are the same as for suprafacial-suprafacial cyclo-additions (see C3.2, page 48), and these two modes of reaction are therefore in competition. However, the suprafacial–suprafacial mode is generally preferred since this mode leads to more effective orbital overlap.

4 Sigmatropic migration reactions

Introduction

Sigmatropic migration reactions involve the migration of an atom or group from an initial point of attachment in a molecule, across an unsaturated system, to a new point of attachment. Common examples include 1,2-alkyl shifts in carbonium ions, and the Cope and Claisen rearrangements. We can represent this type of reaction in general terms by:

$$\underset{\text{C—(C=C)}_n}{\overset{\text{R}}{|}} \longrightarrow \text{C}\overset{\text{R}}{\cdots}(\text{C})_{2n-1}\cdots\text{C} \longrightarrow (\text{C=C})_n\text{—}\overset{\text{R}}{\underset{|}{\text{C}}}$$

The reactant and product usually contain no element of symmetry which is of use in an analysis for conservation of orbital symmetry. The construction of orbital correlation diagrams is therefore impossible without detailed molecular orbital calculations. However, the transition state in a sigmatropic migration usually does contain useful symmetry, and this is therefore used in a *frontier orbital analysis* for conservation of orbital symmetry.

For a concerted migration of R (bond cleavage and bond formation occurring simultaneously), the orbitals, which hold R and the unsaturated system over which it is migrating together, must have the correct symmetry to allow continuous bond cleavage and bond formation.

S4.1 In the Frontier Orbital analysis, we picture the formation of the transition state to involve partial cleavage of the C–R bond, one electron in that bond becoming associated with R and the other with the unsaturated system. The representation of the migration as an R radical migrating over a C⋯$(C)_{2n-1}$⋯C radical is a simplified yet reasonable way of describing the molecular orbitals of the changing system, but we must remember that R does not become completely detached from the unsaturated system, and that in reality both bonding electrons are delocalised over the whole system.

The important point is, which orbitals of R and C⋯$(C)_{2n-1}$⋯C contain the two bonding electrons. For the moment, we will consider R to be a hydrogen atom, and therefore any electron associated with H will be in a 1s orbital. In the initial reactant there were $2n$ π-electrons, so in the unsaturated system of the transition state there will be $2n + 1$ π-electrons, and the $(2n + 1)$th electron will be in the $(n + 1)$th molecular orbital of the C⋯$(C)_{2n-1}$⋯C system; i.e. the highest occupied molecular orbital (H.O.M.O.).

	Q4.1 For the C⋯$(C)_3$⋯C radical in its ground state, which orbital is the highest occupied molecular orbital (H.O.M.O.)?
A4.1 ψ_3 is the H.O.M.O. Electronic configuration: $\psi_1^2\psi_2^2\psi_3^1$	**Q4.2** For the C⋯$(C)_1$⋯C radical in its ground state, which orbital is the H.O.M.O.?
A4.2 ψ_2 is the H.O.M.O. Electronic configuration: $\psi_1^2\psi_2^1$	**Q4.3** For the C⋯$(C)_2$⋯C radical cation in its ground-state, which orbital is the H.O.M.O.? [Note: this system contains an even number of carbon atoms and therefore does not belong to the C⋯$(C)_{2n-1}$⋯C group discussed in S4.1. For an even atom system, C⋯$(C)_{2n-2}$⋯C (n = 1, 2, 3, . . .), in its ground-state, the H.O.M.O. for the radical cation will be the nth molecular orbital, and for the radical anion will be the $(n + 1)$th molecular orbital].

A4.3 ψ_2 is the H.O.M.O.
Electronic configuration: $\psi_1^2\psi_2^1$

S4.2 A hydrogen atom at $C_{(5)}$ of *cis*-penta-1,3-diene will only migrate to $C_{(1)}$ in a concerted manner if the 1s orbital of the hydrogen and the H.O.M.O. of the C==(C)$_3$==C radical have the same symmetry (with respect to the symmetry element present in the transition state).

$$\underset{\large CH_2-CH=CH-CH=CH_2}{\overset{H}{|}} \longrightarrow CH_2{\cdots}CH{\cdots}CH{\cdots}CH{\cdots}CH_2 \text{ (H bridging)} \longrightarrow CH_2=CH-CH=CH-\overset{H}{\overset{|}{CH_2}}$$

The bonding orbital of the transition state may be represented as:

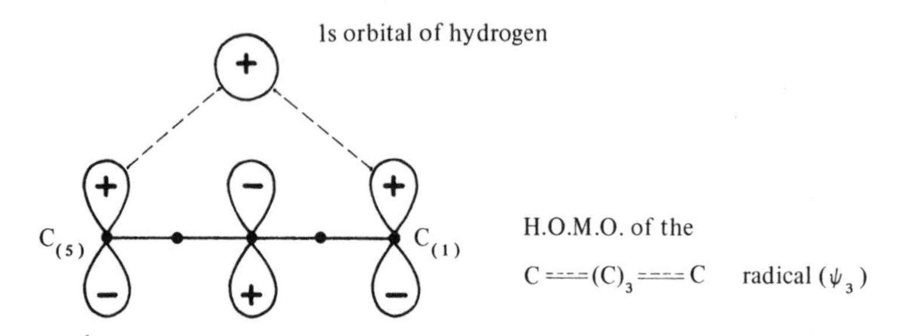

Notice that the interaction between the 1s orbital of the hydrogen atom and the components of ψ_3 of the C==(C)$_3$==C radical at both $C_{(5)}$ and $C_{(1)}$ is a bonding one. Thus, as the $C_{(5)}$–H bond breaks, the $C_{(1)}$–H bond can form in a continuous manner. We are, in fact, describing the bonding orbital of the transition state as a linear combination of 1s and ψ_3. Notice, too, that the hydrogen remains on one face of the unsaturated system (this is a *suprafacial*[1, 5] *migration*), and that the transition state has a plane of symmetry passing through $C_{(3)}$ and the migrating hydrogen (both ψ_3 and the s-orbital are symmetric with respect to this plane).

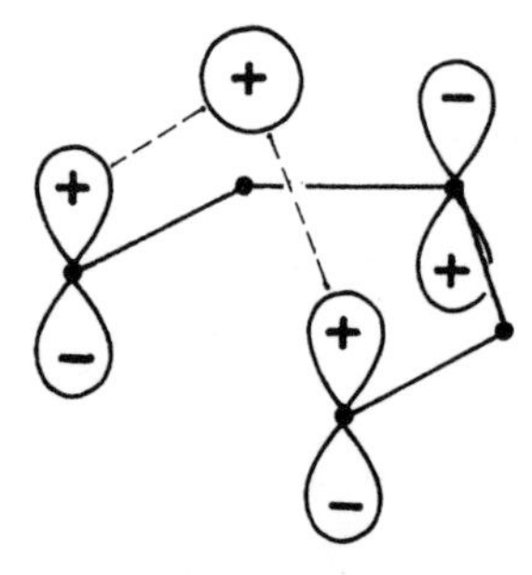

Suprafacial [1,5] migration of hydrogen.

Q4.4 Draw the orbital interactions for a suprafacial-[1, 2] migration of hydrogen in a carbonium ion:

$$\overset{H}{\overset{|}{C}}-C^+ \rightarrow C \overset{H}{\overset{+}{\cdots}} C \rightarrow {}^+C-\overset{H}{\overset{|}{C}}$$

A4.4 The H.O.M.O. of ethylene radical cation is ψ_1.

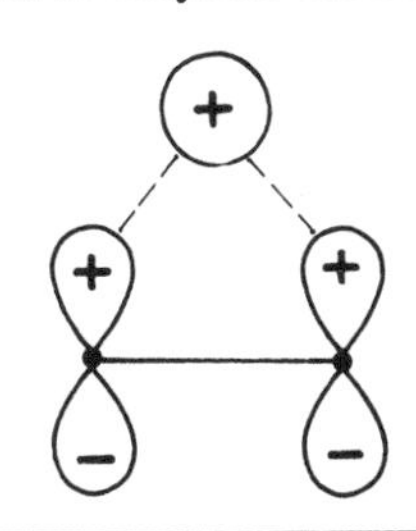

Q4.5 Draw the orbital interactions for a suprafacial [1, 3] migration of hydrogen:

$$\overset{\displaystyle H}{\overset{|}{C}}\text{—C=C} \rightarrow \text{C}\overset{H}{\text{---C---}}\text{C} \rightarrow \text{C=C—}\overset{\displaystyle H}{\overset{|}{C}}$$

A4.5 The H.O.M.O. of a prop-2-enyl radical is ψ_2.

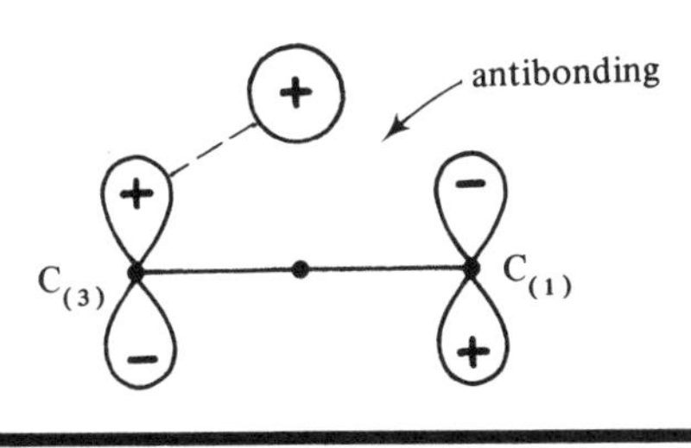

C4.1 A suprafacial [1, 3] migration of hydrogen is symmetry forbidden. Although the interaction between the hydrogen and $C_{(3)}$ is a bonding one, there is an antibonding interaction between the hydrogen and $C_{(1)}$. Bond cleavage and bond formation cannot occur in a concerted manner. Notice that the 1s orbital of the hydrogen and the ψ_2 orbital of the prop-2-enyl system have different symmetry classifications with respect to the plane of symmetry present in a suprafacial migration.

S4.3 [1, 3] Migration of hydrogen can occur in a concerted manner if the hydrogen migrates to the opposite face of the unsaturated system (an *antarafacial* [1, 3] *migration* of hydrogen).

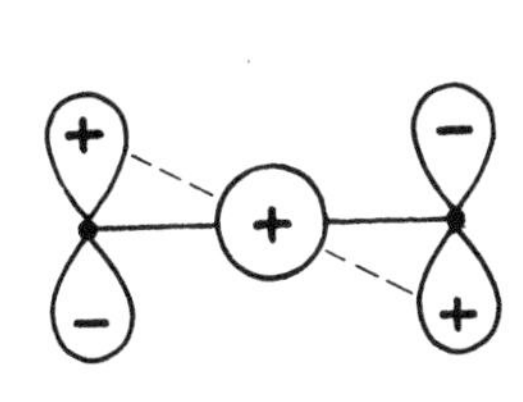

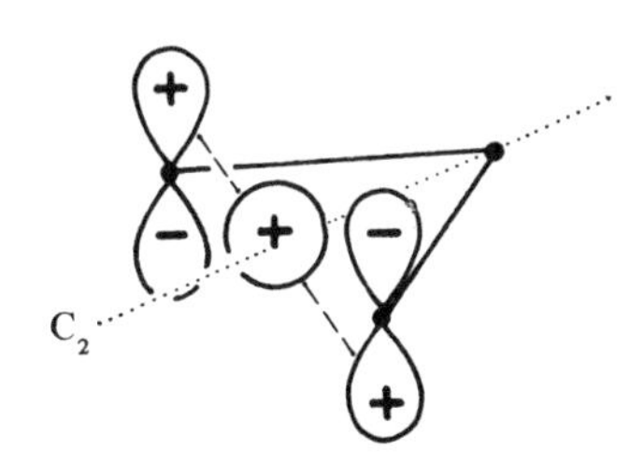

However, the lack of flexibility in such a short unsaturated system prevents efficient overlap of the orbitals involved. No examples are known. Notice that for an antarafacial migration there exists a 2-fold axis of symmetry.

Q4.6 Is the concerted, thermal [1,7] migration of hydrogen likely to occur by a suprafacial or an antarafacial route?

A4.6 The H.O.M.O. of a C⩦(C)$_5$⩦C radical is ψ_4.

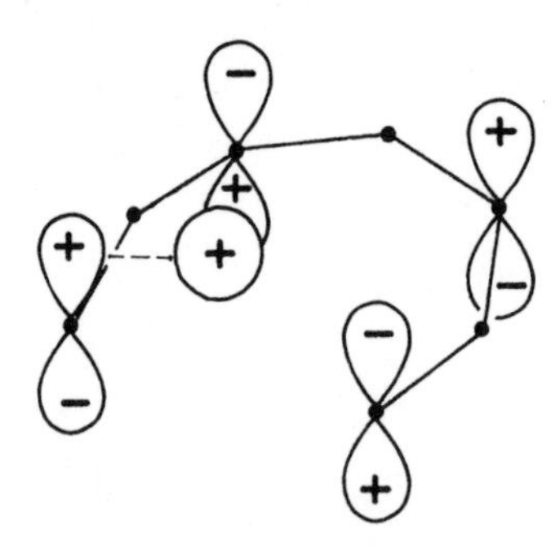

Suprafacial [1,7] migration of hydrogen-symmetry forbidden

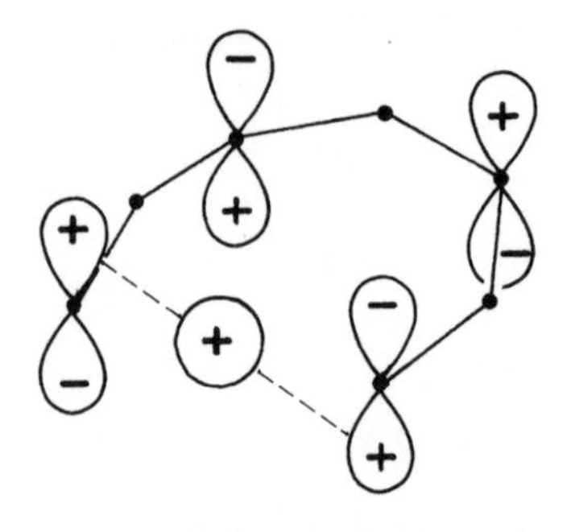

Antarafacial [1,7] migration of hydrogen-symmetry allowed

[1,7] Migration of hydrogen is likely to occur by an antarafacial route. Examples are known.

Q4.7 Is a concerted, thermal [1,4] migration of hydrogen in a but-2-enyl anion likely to occur by a suprafacial or an antarafacial route?

A4.7 The H.O.M.O. of a C⩦(C)$_2$⩦C radical anion is ψ_3.

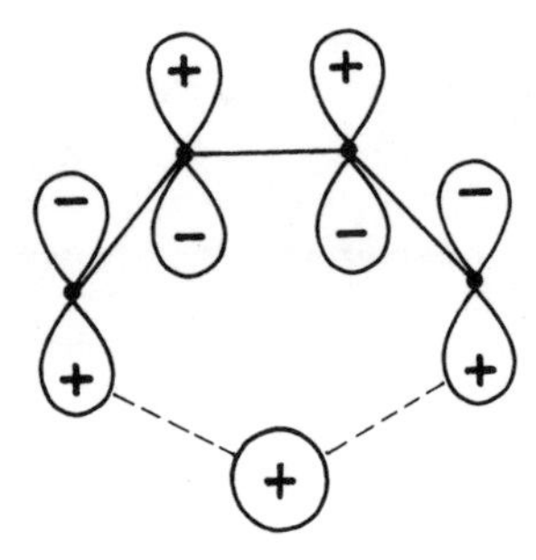

Suprafacial [1,4] migration of hydrogen-symmetry allowed (antarafacial [1,4] migration of hydrogen-symmetry forbidden).

[1,4] Migration of hydrogen in an anion should occur by a suprafacial route.

S4.4 Sigmatropic migrations occurring from the 1st excited-state of the reactant will involve a different highest occupied molecular orbital in the transition state to that involved in the thermal reaction. This will be the orbital above the H.O.M.O. for the thermal reaction; e.g. for a photochemical [1,3] migration of hydrogen, the H.O.M.O. will be ψ_3 of the prop-2-enyl system.

Q4.8 For a concerted, photochemical [1,3] migration of hydrogen, is the symmetry allowed route suprafacial or antarafacial?

A4.8 The H.O.M.O. for a photochemical [1,3] migration is ψ_3.

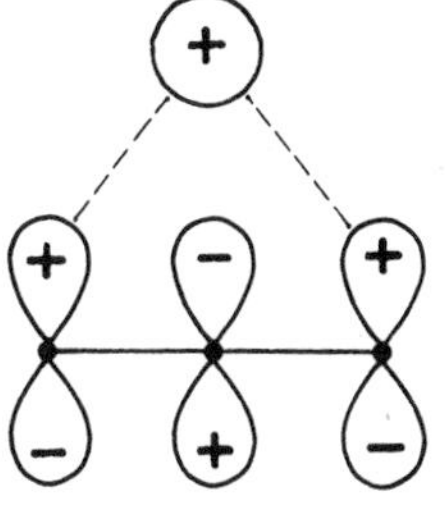

The symmetry allowed route is suprafacial

Q4.9 For a concerted, photochemical [1,4] migration of hydrogen in a but-2-enyl cation, is the symmetry allowed route suprafacial or antarafacial?

A4.9 The H.O.M.O. for a photochemical [1,4] migration in a cation is ψ_3.

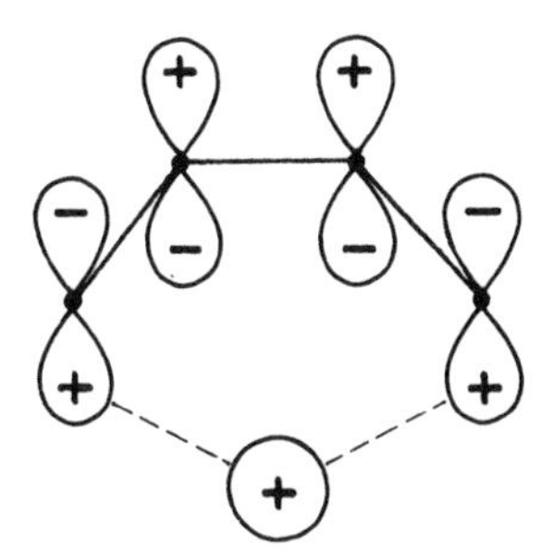

The symmetry allowed route is suprafacial

C4.2 *Stereochemical course for the concerted migration of hydrogen*

Migration	Thermal	Photochemical
[1,3]	Antarafacial•	Suprafacial
[1,5]	Suprafacial	Antarafacial•
[1,7]	Antarafacial	Suprafacial
[1,2] Cation	Suprafacial	Antarafacial•
[1,2] Anion	Antarafacial•	Suprafacial
[1,4] Cation	Antarafacial•	Suprafacial
[1,4] Anion	Suprafacial	Antarafacial•

• Antarafacial modes in [1,2], [1,3], [1,4] and probably also [1,5] migrations are unlikely to occur due to inefficient orbital overlap.

Q4.10 At 150°, *A* undergoes an uncatalysed hydrogen migration to give *B*.

(*A*) ⇌ (*B*)

This could occur by several pathways: (a) two consecutive suprafacial [1,3] shifts of hydrogen, (b) two consecutive antarafacial [1,3] shifts of hydrogen, (c) one suprafacial and one antarafacial [1,3] shift of hydrogen, (d) one suprafacial [1,5] shift of hydrogen, and (e) one antarafacial [1,5] shift of hydrogen. Which routes involve conservation of orbital symmetry, and which is the most probable route?

A4.10 Routes (b) and (d) involve conservation of orbital symmetry; routes (a), (c) and (e) do not. The most probable route is (d), a suprafacial [1,5] shift of hydrogen.

(D. S. Glass, R. S. Boikess and S. Winstein, *Tetrahedron Letters*, 1966, 999.)

Q4.11 Thermal [1,5] migration of hydrogen in *A* gives *B* and *C*. Predict the stereochemistry of the carbon atoms marked (*) in the products *B* and *C*.

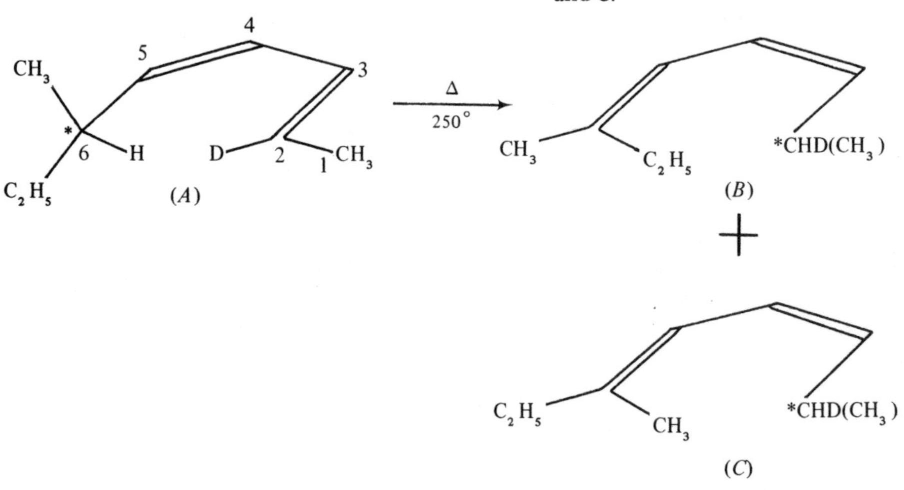

A4.11 For a concerted, thermal [1,5] migration of hydrogen, the pathway must be suprafacial. By rotation of $C_{(5)}—C_{(6)}$ in *A*, place *A* in the two conformations which will allow suprafacial migration of hydrogen from $C_{(6)}$. As the hydrogen migrates, the orientation of the other groups at $C_{(6)}$ will become fixed by the formation of the $C_{(5)}C_{(6)}$ double bond. The stereochemistry of the two products will be:

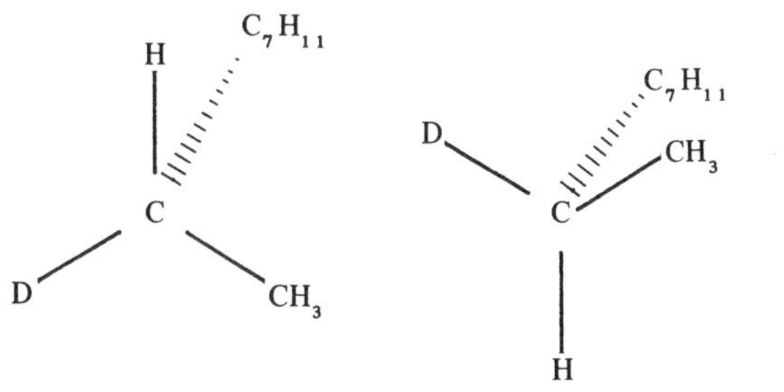

(*B*) (*C*)

(W. R. Roth, J. Konig, and K. Stein, *Chem. Ber.*, 1970, 103, 426.)

S4.5 Elements which use p- and d-orbitals for bonding do not have to follow the same stereochemical course as predicted for hydrogen. For example, although the concerted, thermal [1,3] migration of hydrogen must occur antarafacially, the concerted, thermal [1,3] migration of carbon can occur suprafacially because carbon can use both lobes of a p-orbital for continuous bonding:

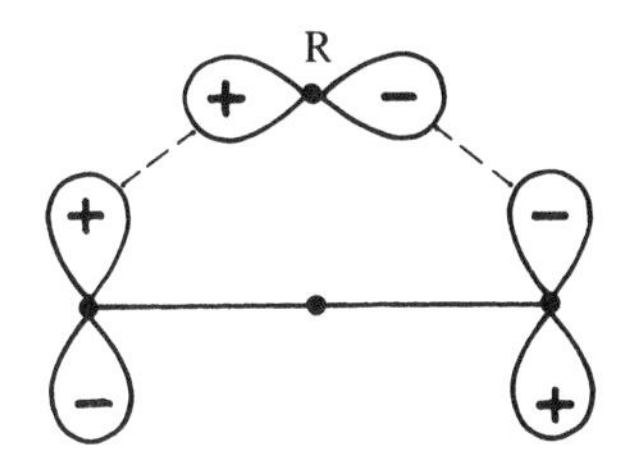

Notice that both orbitals are antisymmetric with respect to the plane of symmetry present in a suprafacial migration.

Since the new bond is formed on the opposite face of R to the bond which is broken, the migration involves an inversion of configuration at R:

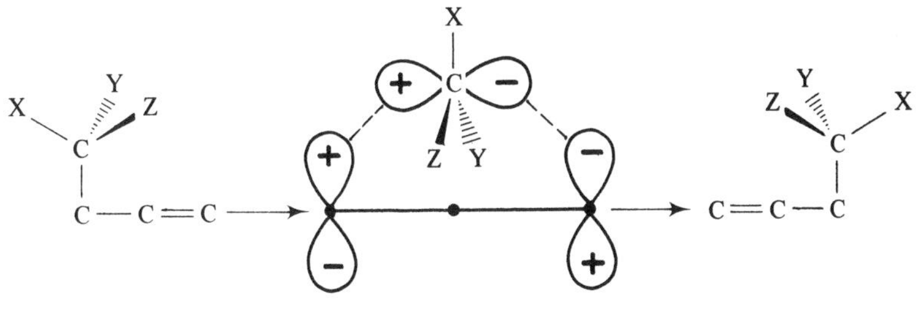

The [1,5] migration of carbon can follow the same course as for hydrogen (suprafacial), in which case only one face of R is involved in bonding. This gives retention of configuration at R:

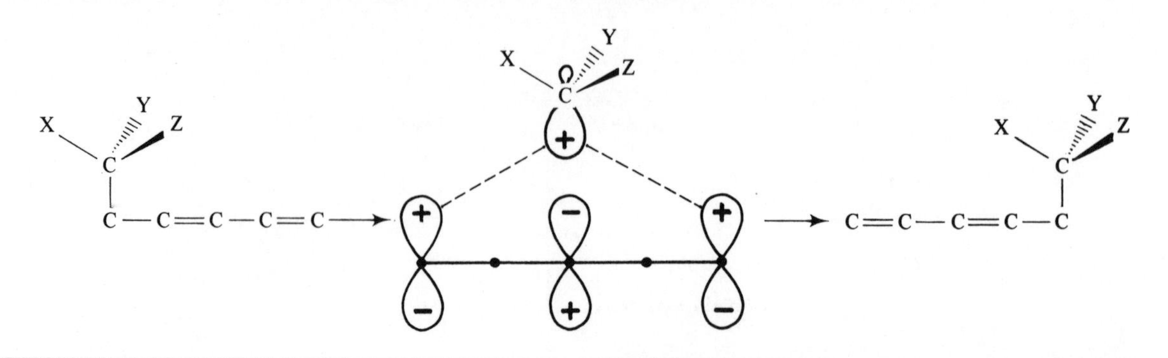

Q4.12 Compared with the original configuration of R, what would be the final configuration of R after (a) a thermal [1,7] suprafacial migration of carbon and (b) a thermal [1,7] antarafacial migration of carbon?

A4.12 The H.O.M.O. for a thermal [1,7] migration is ψ_4.

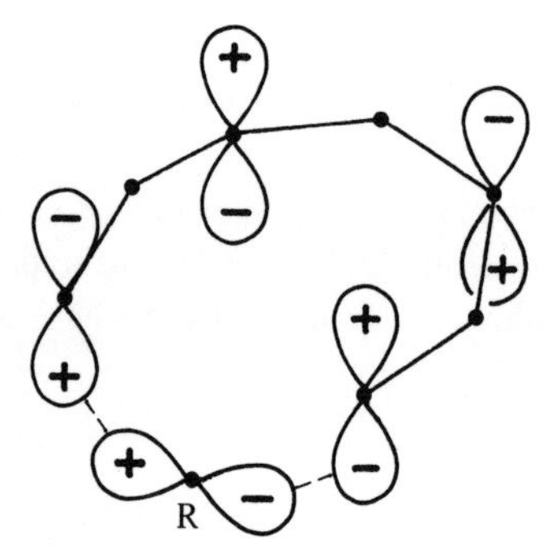

(a) Suprafacial migration-inversion of configuration

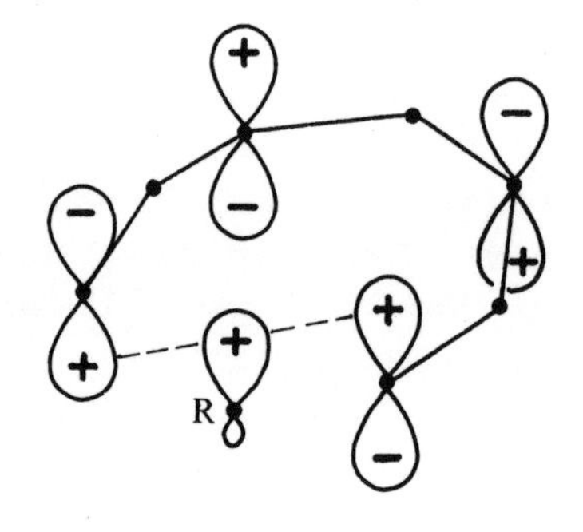

(b) Antarafacial migration-retention of configuration

Q4.13 Predict the stereochemical course for a concerted, thermal [1,2] migration of carbon in (a) a cation, and (b) an anion.

A4.13 (a) The H.O.M.O. for a thermal [1,2] migration in a cation is ψ_1.

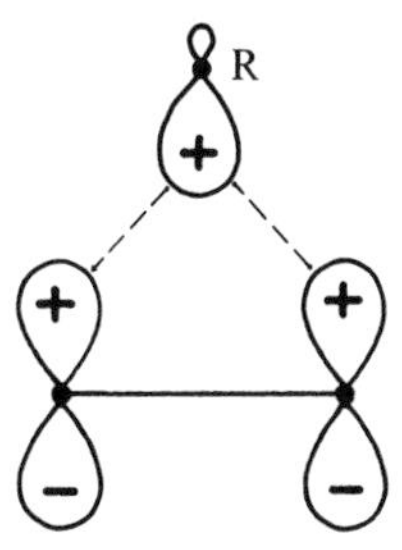

Retention of configuration at R in a suprafacial migration. Antarafacial migration unlikely

(b) The H.O.M.O. for a thermal [1,2] migration in an anion is ψ_2.

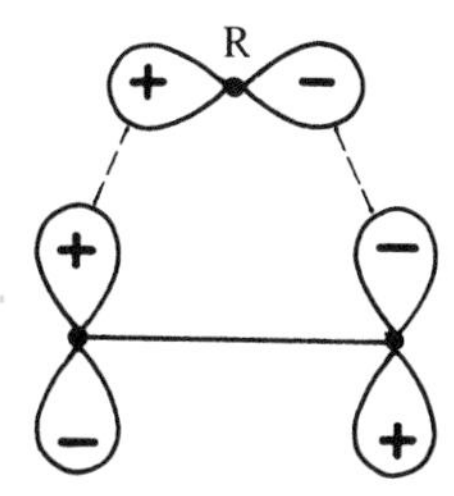

Inversion of configuration at R in a suprafacial migration. Antarafacial migration unlikely

Q4.14 For a concerted, thermal [1,4] suprafacial migration of carbon in (a) a cation, and (b) an anion, what will be the final configuration of the migrating group?

A4.14 (a) The H.O.M.O. for a thermal [1,4] migration in a cation is ψ_2.

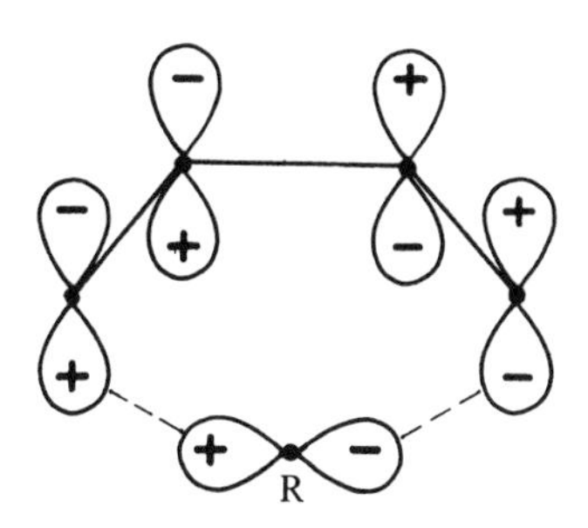

Inversion of configuration at R

(b) The H.O.M.O. for a thermal [1,4] migration in an anion is ψ_3.

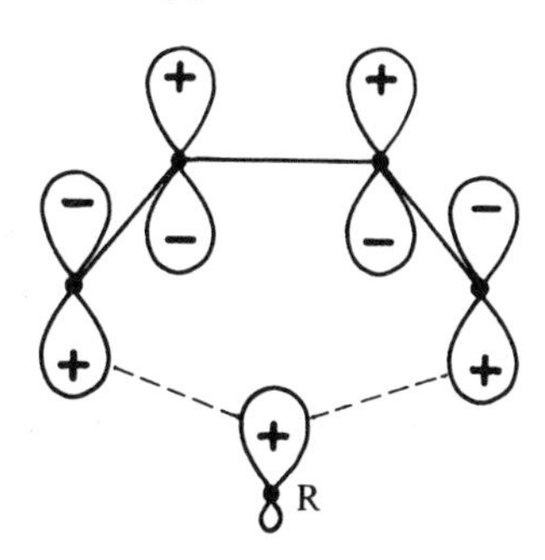

Retention of configuration at R

Q4.15 Predict the stereochemical course for a concerted, photochemical [1,3] migration of carbon.

A4.15 The H.O.M.O. for a photochemical [1,3] migration is ψ_3.

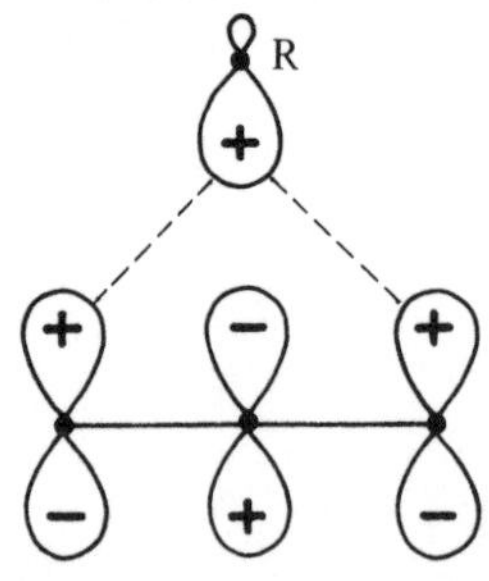

Retention of configuration at R in a suprafacial migration. Antarafacial migration unlikely

Q4.16 For a concerted, photochemical [1,5] suprafacial migration of carbon, what will be the final stereochemistry of the migrating group?

A4.16 The H.O.M.O. for a photochemical [1,5] migration is ψ_4.

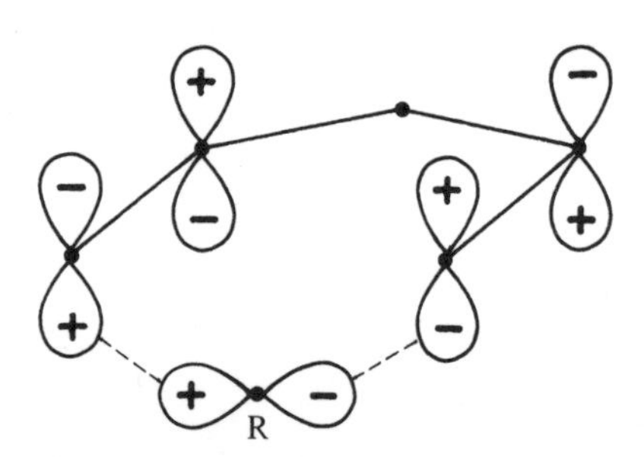

Inversion of configuration at R

C4.3 *Stereochemical course for the concerted migration of carbon*

Migration		Thermal	Photochemical
[1,3] →	Suprafacial	Inversion	Retention
	Antarafacial•	Retention	Inversion
[1,5] →	Suprafacial	Retention	Inversion
	Antarafacial•	Inversion	Retention
[1,7] →	Suprafacial	Inversion	Retention
	Antarafacial	Retention	Inversion
[1,2] cation →	Suprafacial	Retention	Inversion
	Antarafacial•	Inversion	Retention

(cont.)

Stereochemical course for the concerted migration of carbon (contd.)

Migration		Mode	
[1,2] anion	Inversion	Suprafacial	Retention
	Retention	Antarafacial•	Inversion
[1,4] cation	Inversion	Suprafacial	Retention
	Retention	Antarafacial•	Inversion
[1,4] anion	Retention	Suprafacial	Inversion
	Inversion	Antarafacial•	Retention

• Antarafacial modes in [1,2], [1,3], [1,4], and probably also [1,5] migrations are unlikely to occur due to inefficient orbital overlap.

Q4.17 Thermal rearrangement of *A* gives a mixture of *B* and *C*, one of which predominates over the other. If the rearrangement occurs with conservation of orbital symmetry, will there be retention or inversion of configuration at the carbon marked (*), and which would you expect to be the major product?

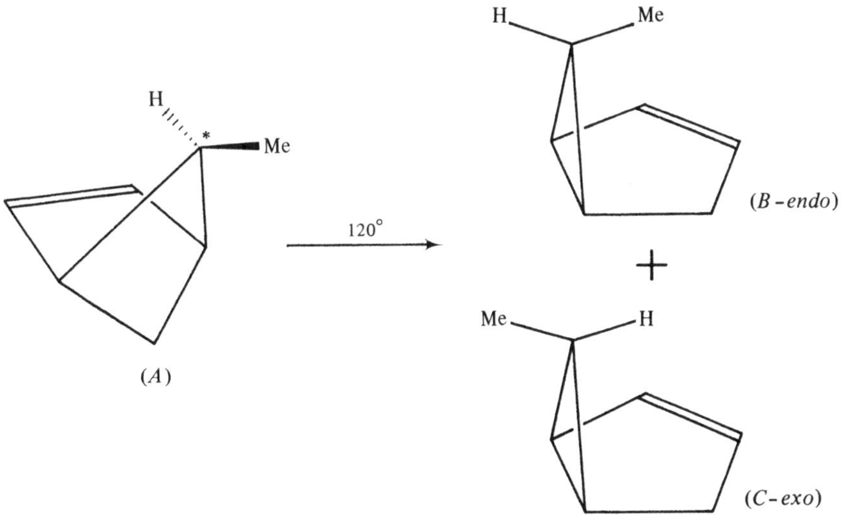

H Me *

(*A*)

120°

H Me

(*B*-*endo*)

+

Me H

(*C*-*exo*)

A4.17 The rearrangement is forced by the molecular geometry to be a suprafacial [1,3] migration of carbon. This must involve *inversion* of configuration at the migrating carbon. Product *B* is formed with retention of configuration at C*, while *C* is formed with inversion of configuration at C*. From conservation of orbital symmetry considerations *C* would be the major product. ($C/B \simeq 200$).

(W. R. Roth and A. Friedrich, *Tetrahedron Letters*, 1969, 2607.)

Q4.18 The cation *A* undergoes a thermal, suprafacial [1,4] migration of $C_{(6)}$. This occurs so rapidly at $-50°$ in $SO_2ClF-FSO_3H$, that the n.m.r. spectrum shows a single, sharp line for the five methyl groups at $C_{(1)}$ to $C_{(5)}$; i.e. the absorption due to these five methyl groups is averaged. Of the two methyl groups at $C_{(6)}$ in *A*, one lies above the five-membered ring, while the other is directed outwards.

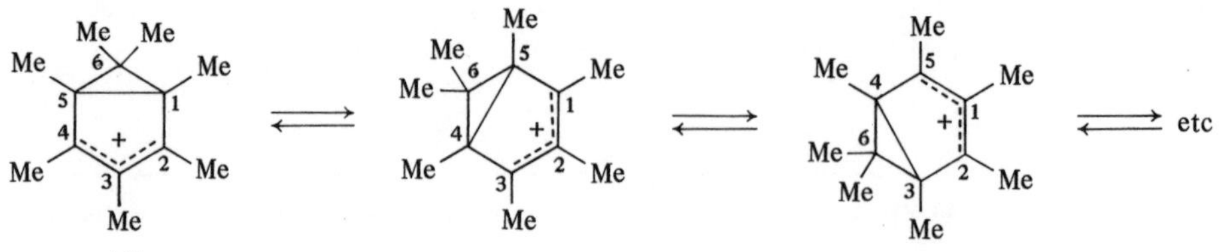

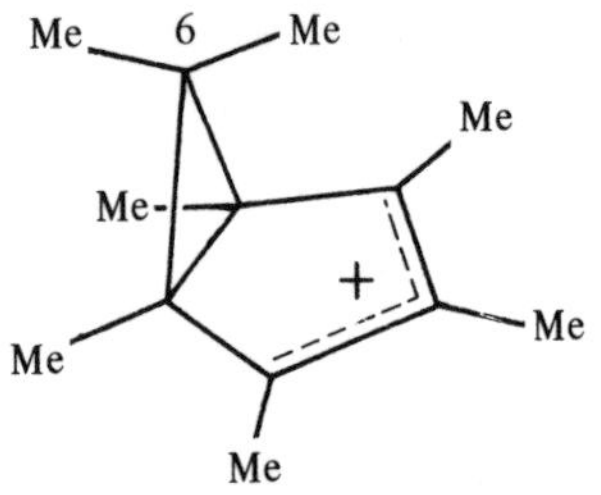

If the rearrangement is concerted, (a) would it involve retention or inversion of configuration at $C_{(6)}$ during each migration step, (b) would this result in interchanging the positions of the two methyl groups at $C_{(6)}$, or would one remain above the five-membered ring and the other remain pointing outwards, and (c) as a result of this, would you expect to obtain two distinct absorptions for the methyl groups at $C_{(6)}$ in the n.m.r. spectrum, or would their absorption be averaged, giving a single, sharp line?

A4.18 (a) A thermal, suprafacial [1,4] migration of carbon in a cation, if concerted, must involve inversion of configuration at the migrating carbon. Therefore, inversion at $C_{(6)}$ must occur during each migration step.

(b) As a result of this, one methyl at $C_{(6)}$ will remain above the five-membered ring, and the other will remain pointing outwards. Their positions will not interchange.

(c) We would expect two distinct absorptions for the methyl groups at $C_{(6)}$, because the environment of each remains unchanged throughout the rearrangement.

(R. F. Childs and S. Winstein, *J. Am. chem. Soc.*, 1968, 90, 7146.)

S 4.6 The migrating group R may also be a delocalised system; e.g. the Cope rearrangement:

1, 2, 3, 1′, 2′, 3′

Here again, the bonding orbital of the transition state is pictured as a linear combination of the highest occupied molecular orbitals of the two fragments.

Q4.19 In the thermal Cope rearrangement, which is the H.O.M.O. of the two prop-2-enyl fragments?

A4.19 ψ_2 of both fragments.

Q4.20 Does the H.O.M.O. of the two prop-2-enyl fragments have the correct symmetry to allow continuous bond cleavage and bond formation? Draw the interactions of the two H.O.M.O.s.

A4.20

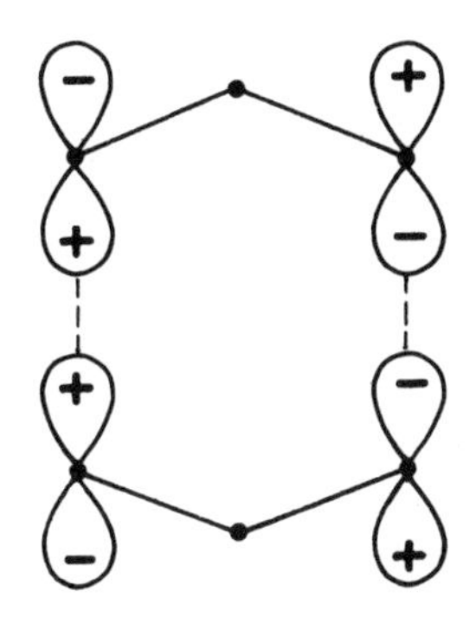

Yes; bond cleavage and bond formation can occur in a concerted manner. The stereochemistry of the change will be suprafacial on both fragments. This is therefore a [3,3] suprafacial-suprafacial migration (the new σ-bond is joined to the third atom from the original σ-bond in both fragments).

Q4.21 Consider both the suprafacial-suprafacial, and the suprafacial-antarafacial modes of reaction for a concerted, thermal [3,5] migration:

2′ 3′ C=C, 1′ C, 1 C, C=C–C=C 2 3 4 5

Which will be the symmetry allowed course for the reaction?

A4.21 The H.O.M.O. for the C=(C)$_1$$\text{=}$C fragment is ψ_2.
The H.O.M.O. for the C=(C)$_3$$\text{=}$C fragment is ψ_3.

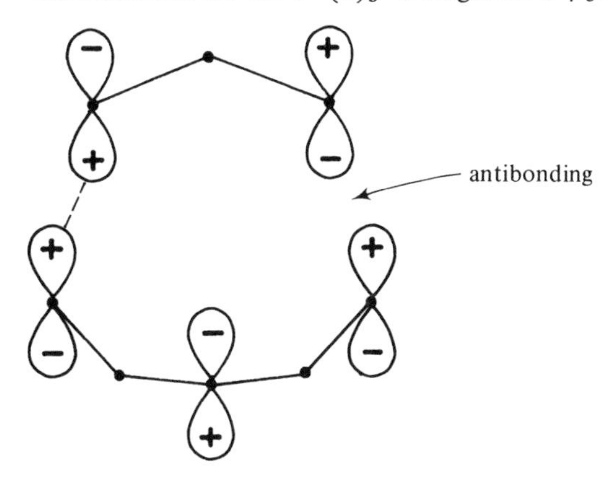

Suprafacial–suprafacial: symmetry forbidden

(cont.)

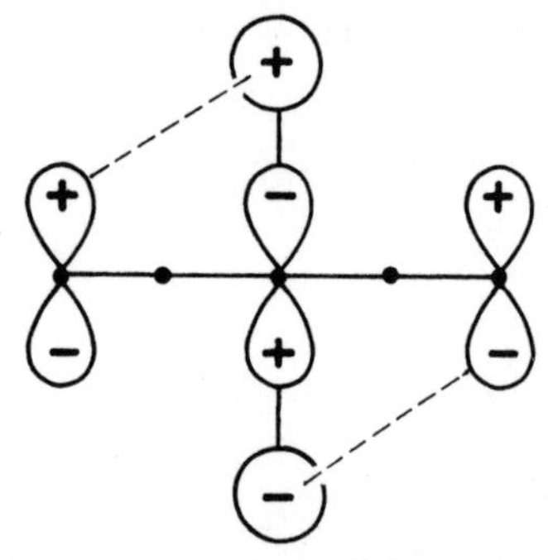

Suprafacial on the C_3 fragment, antarafacial on the C_5 fragment: symmetry allowed.

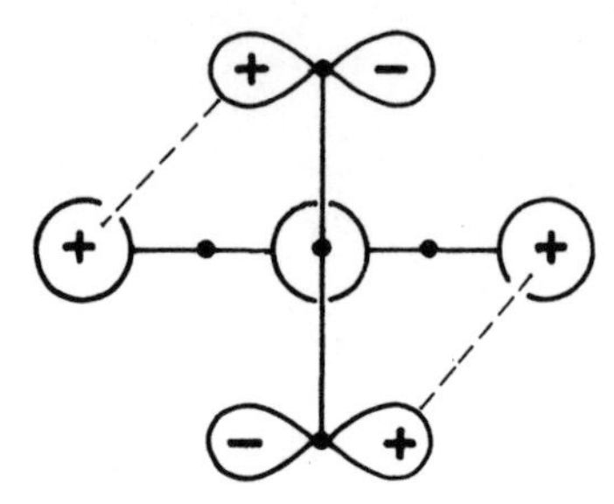

Suprafacial on the C_5 fragment, antarafacial on the C_3 fragment: symmetry allowed.

Both suprafacial-antarafacial modes are symmetry allowed.

5 State correlation diagrams

Introduction

The orbital correlations which we used to analyse both electrocyclic reactions and cyclo-addition reactions still give a somewhat simplified picture of the reacting system as a whole, and some of the orbital correlations which we predicted may not in fact occur. The cause of this is electron repulsion which is ignored in deriving the molecular orbitals for the systems we have analysed. In reality, electron repulsion *is* present, and in some cases prevents the orbital correlations which we would have expected on the basis of a simple analysis. These 'non-correlations' are brought to light by more refined molecular orbital calculations on species whose geometry lies between that of the reactant(s) and product(s).

Nevertheless, the use of orbital correlation diagrams to predict whether a particular reaction will be symmetry allowed or forbidden does give correct results, even though some of the details may be oversimplified.

We can obtain a better prediction of the correlation of the electronic configurations of the reactant(s) and product(s) by the use of *state correlation diagrams.*• These are constructed in a similar manner to orbital correlation diagrams, but now we must use the total symmetry of all the electrons involved in the reaction with respect to the element of symmetry present in the reaction. This is known as the *state symmetry,* state referring to the electronic state or configuration.

S5.1 The symmetry of a particular electronic state is obtained by multiplying together the symmetries of the occupied orbitals for *each* electron. This is analogous to the multiplication of positive and negative numbers, with positive corresponding to symmetric and negative to antisymmetric orbitals or states. Consider a conrotatory electrocyclic reaction of buta-1,3-diene in its ground-state, in which there is a 2-fold axis of symmetry. The electronic state (configuration) is $\psi_1^2\psi_2^2$, and the symmetry of these orbitals with respect to the 2-fold axis of symmetry is ψ_1(A) and ψ_2(S). The *state symmetry* is therefore:

$$\psi_1^1\ \psi_1^1\ \psi_2^1\ \psi_2^1 = \psi_1^2\psi_2^2$$
(A)(A)(S) (S) = (S) State symmetry
(−)(−)(+) (+) = (+)

Q5.1 What is the state symmetry of the 1st excited-state of buta-1,3-diene in a conrotatory electrocyclic reaction?

A5.1 Electronic state: $\psi_1^2\psi_2^1\psi_3^1$
Symmetry element: C_2
Symmetry of orbitals: ψ_1 (A), ψ_2 (S), ψ_3 (A)
State symmetry: $\psi_1^1\ \psi_1^1\ \psi_2^1\ \psi_3^1 = \psi_1^2\psi_2^1\psi_3^1$
(A)(A)(S) (A) = (A) State symmetry
(−)(−)(+) (−) = (−)

• The use of both orbital and state correlation diagrams in an analysis for conservation of orbital symmetry was introduced by H. C. Longuet-Higgins and E. W. Abrahamson (*J. Am. chem. Soc.*, 1965, 87, 2045).

Q5.2 What is the state symmetry of the ground-state of buta-1,3-diene in a disrotatory electrocyclic reaction?

A5.2 Electronic state: $\psi_1^2\psi_2^2$
Symmetry element: m
Symmetry of orbitals: ψ_1 (S), ψ_2 (A)
State symmetry: $\psi_1^1\ \psi_1^1\ \psi_2^1\ \psi_2^1 = \psi_1^2\psi_2^2$
(S) (S) (A)(A) = (S) State symmetry
(+) (+) (−)(−) = (+)

Q5.3 What is the state symmetry of the 1st excited-state of buta-1,3-diene in a disrotatory electrocyclic reaction?

A5.3 Electronic state: $\psi_1^2\psi_2^1\psi_3^1$
Symmetry element: m
Symmetry of orbitals: ψ_1 (S), ψ_2 (A), ψ_3 (S)
State symmetry: $\psi_1^1\ \psi_1^1\ \psi_2^1\ \psi_3^1 = \psi_1^2\psi_2^1\psi_3^1$
(S) (S) (A)(S) = (A)

Q5.4 What is the state symmetry of the ground-state of cyclobutene in (a) a conrotatory, and (b) a disrotatory, electrocyclic opening to buta-1,3-diene?

A5.4 Electronic state: $\sigma^2\pi^2$
(a) Symmetry element: C_2
Symmetry of orbitals: σ (S), π (A)
State symmetry: (S) (S) (A) (A) = (S)

(b) Symmetry element: m
Symmetry of orbitals: σ (S), π (S)
State symmetry: (S) (S) (S) (S) = (S)

Q5.5 What is the state symmetry of the 1st excited-state of cyclobutene in (a) a conrotatory, and (b) a disrotatory, electrocyclic opening to buta-1,3-diene?

A5.5 Electronic state: $\sigma^2\pi^1\pi^{*1}$
(a) Symmetry element: C_2
Symmetry of orbitals: σ (S), π (A), π^* (S)
State symmetry: (S) (S) (A) (S) = (A)

(b) Symmetry element: m
Symmetry of orbitals: σ (S), π(S), π^* (A)
State Symmetry: (S) (S) (S) (A) = (A)

S5.2 A state correlation diagram is constructed as follows:

1) arrange the various electronic states of the reactant(s) and product(s) in order of increasing energy, the ground-state at the bottom;

2) beside each electronic state indicate its state symmetry.

For the disrotatory interconversion of buta-1,3-diene and cyclobutene (A2.39, p. 29):

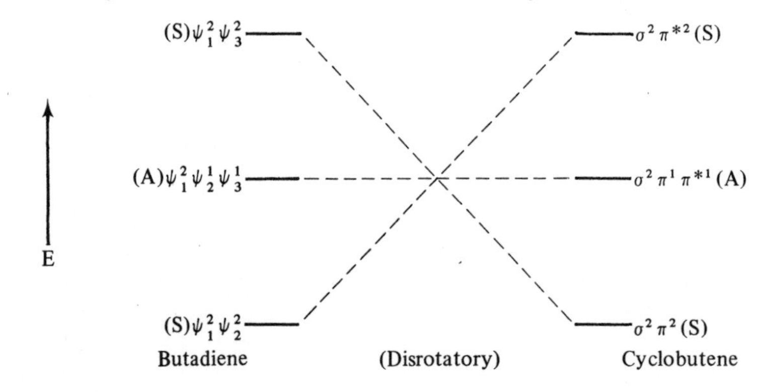

An orbital correlation diagram would predict the correlations shown by the broken lines; i.e. $\psi_1^2\psi_2^2$ (S) would correlate with $\sigma^2\pi^{*2}$ (S), $\sigma^2\pi^2$ (S) with $\psi_1^2\psi_3^2$ (S), and $\psi_1^2\psi_2^1\psi_3^1$ (A) with $\sigma^2\pi^1\pi^{*1}$ (A). However, more refined molecular orbital calculations show that the correlations of $\psi_1^2\psi_2^2$ (S) with $\sigma^2\pi^{*2}$ (S), and $\sigma^2\pi^2$ (S) with $\psi_1^2\psi_3^2$ (S), do not in fact take place. Electron repulsion, ignored in less sophisticated calculations, prevents the lines, correlating the four symmetric (S) states, from crossing. This is known as the *non-crossing rule* of states of the same symmetry. Instead, $\psi_1^2\psi_2^2$ (S) is forced to correlate with $\sigma^2\pi^2$ (S), and $\psi_1^2\psi_3^2$ (S) with $\sigma^2\pi^{*2}$ (S), as shown by the unbroken lines:

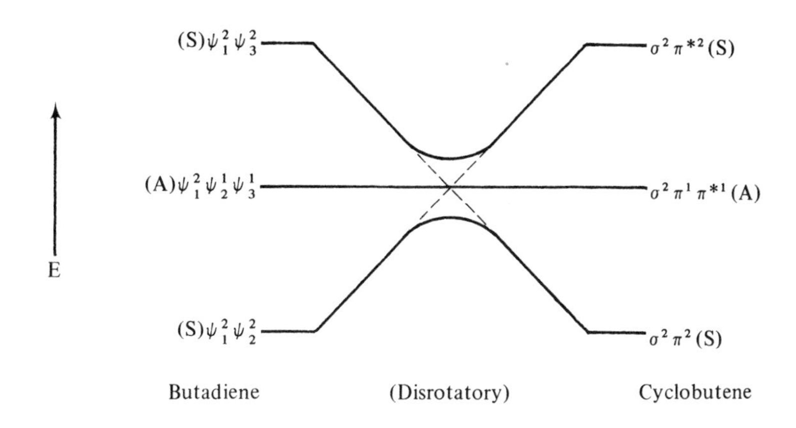

The effect of electron repulsion is not important until the reaction has proceeded some way along the reaction co-ordinate, and until this point is reached the reaction involves an *increase* in energy imposed by the conservation of orbital symmetry. Therefore, although the ground-states of butadiene and cyclobutene do correlate in a disrotatory change, a considerable amount of energy is required to bring about this change.

There is therefore a third rule in constructing state correlation diagrams:

3) draw correlation lines between states of the same symmetry, but lines connecting two pairs of states of the same symmetry must not cross; in the region of the intended crossing point these correlations lines become modified as shown.

There are other electronic states of butadiene and cyclobutene which are not shown in the diagram; these correlate with higher excited-states than those shown. The states $\psi_1^2\psi_3^2$ and $\sigma^2\pi^{*2}$ are included simply because the orbital correlation diagram predicted a correlation of these states with the ground-states of cyclobutene and butadiene respectively.

Q5.6 to 5.11 In the state correlation diagrams required for these questions, include three electronic states only for the reactant(s) and three only for the product(s). These should be: 1) the ground-states, 2) the 1st excited-states, of the reactant(s) and the product(s), and 3) the higher excited-states which are predicted by the appropriate orbital correlation diagram to correlate with the ground-state or the 1st excited-state of the product(s) and the reactant(s) respectively.

Q5.6 Construct a state correlation diagram for the conrotatory interconversion of buta-1,3-diene and cyclobutene (see S2.8, p. 28).

A5.6

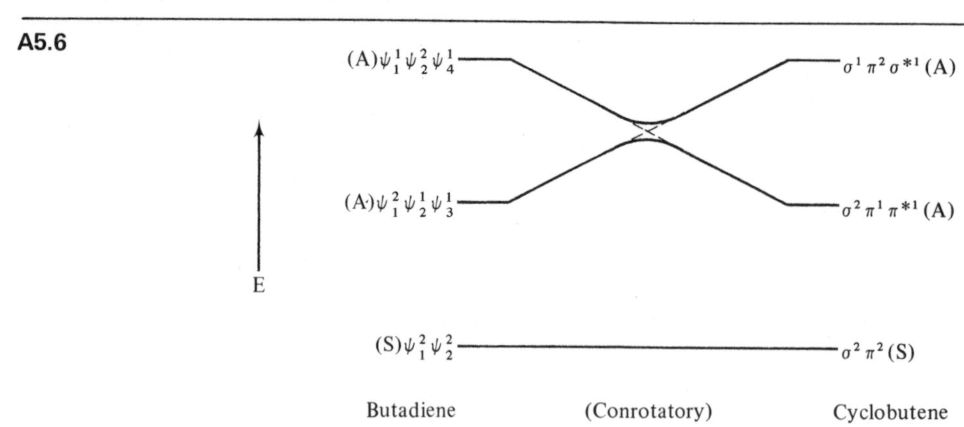

The states $\psi_1^1\psi_2^2\psi_4^1$ and $\sigma^1\pi^2\sigma^{*1}$ are included because the orbital correlation diagram predicted a correlation of these states with the 1st excited-states of cyclobutene and butadiene respectively. Other electronic states have been omitted.

S5.3 We see that the state correlation diagrams for both the conrotatory and the disrotatory modes show a correlation of the ground-state of buta-1,3-diene with the ground-state of cyclobutene. As far as conservation of orbital symmetry is concerned, there is little energy change involved in the conrotatory process, but in the disrotatory process the initial stage of the reaction involves an increase in energy. The conrotatory process is therefore the preferred mode of change for a thermal reaction.

In the photochemical reaction from the 1st excited-state, the disrotatory process is preferred to the conrotatory one because the latter involves an increase in energy in the initial stage of the reaction. There is little energy change involved in the disrotatory process.

Q5.7 Construct a state correlation diagram for the disrotatory interconversion of a prop-2-enyl cation and a cyclopropyl cation (see A2.54, page 33).

A5.7

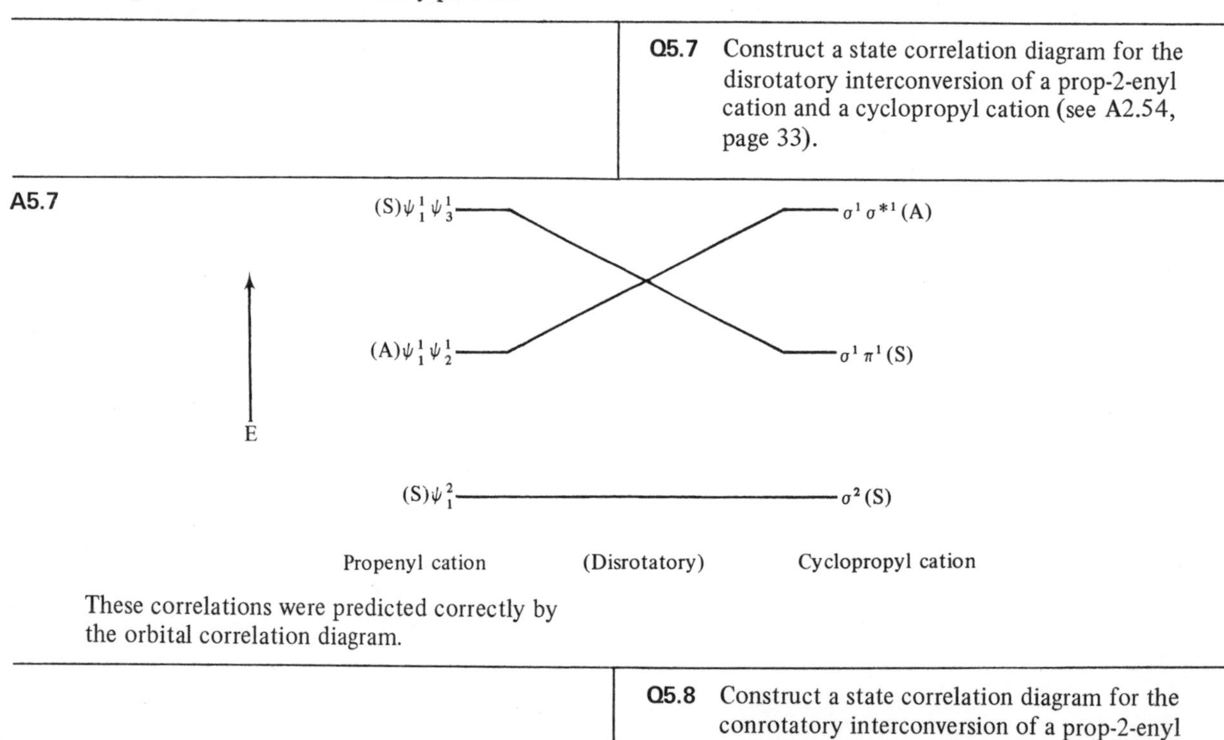

These correlations were predicted correctly by the orbital correlation diagram.

Q5.8 Construct a state correlation diagram for the conrotatory interconversion of a prop-2-enyl cation and a cyclopropyl cation (see A2.53. page 33).

A5.8

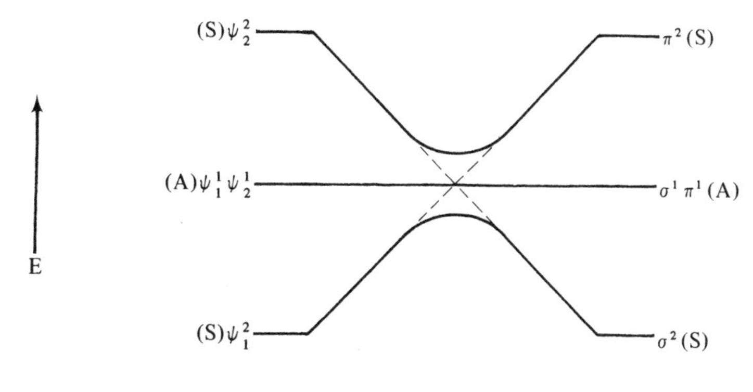

The correlations predicted by the orbital correlation diagram are shown by broken lines.

C5.1 Although the state correlation diagrams for both the disrotatory and the conrotatory interconversion of a prop-2-enyl cation and a cyclopropyl cation show a correlation between the ground-states of both species, the disrotatory process is preferred because in this mode there is no initial energy increase. For the same reason, the conrotatory process is preferred for a reaction involving the 1st excited-state.

Q5.9 Construct a state correlation diagram for the conrotatory interconversion of a prop-2-enyl anion and a cyclopropyl anion (see A2.53, page 33).

A5.9

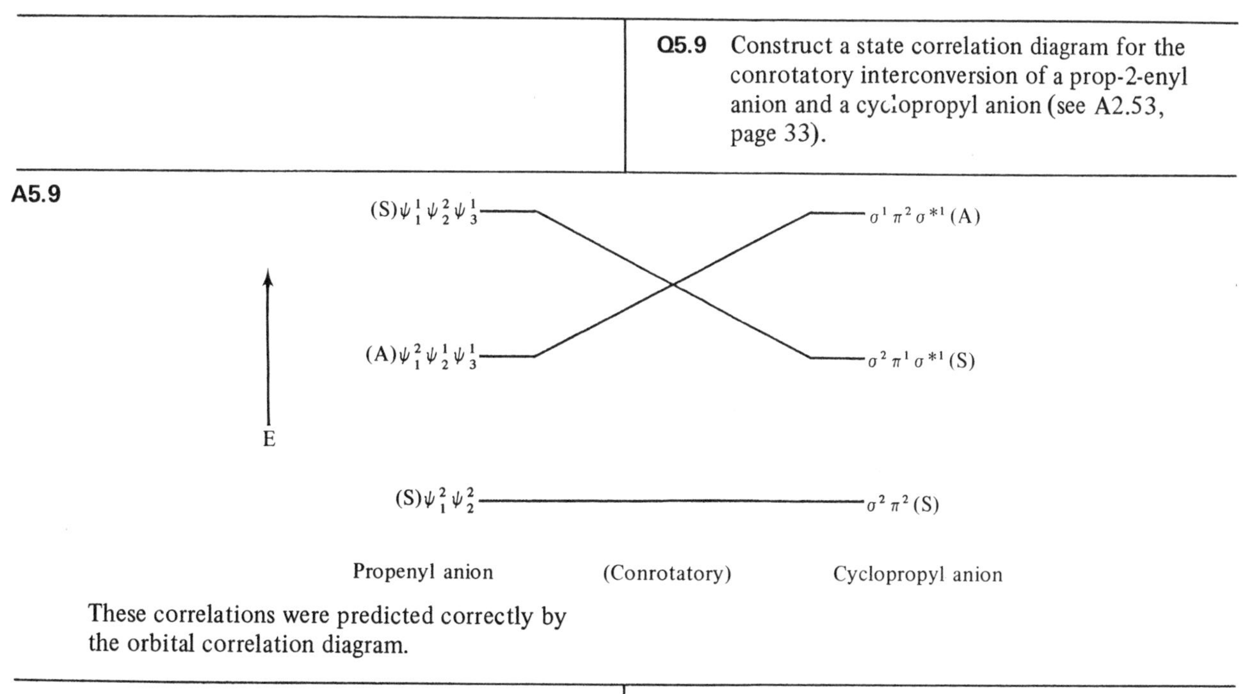

These correlations were predicted correctly by the orbital correlation diagram.

Q5.10 Construct a state correlation diagram for the disrotatory interconversion of a prop-2-enyl anion and a cyclopropyl anion (see A2.54, page 33).

A5.10

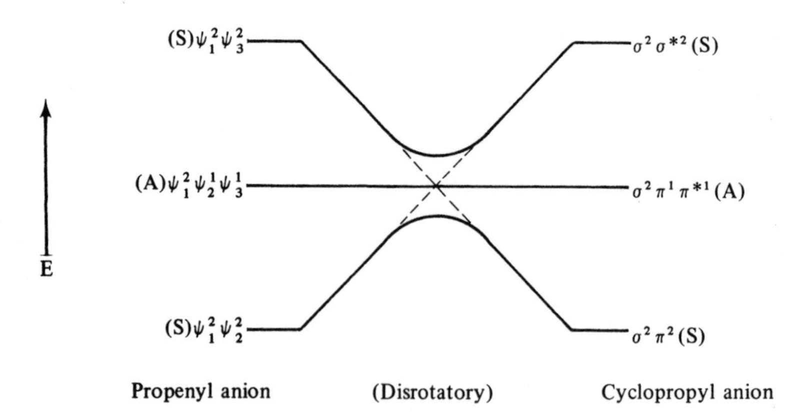

The correlations predicted by the orbital correlation diagram are shown by broken lines.

C5.2 For the interconversion of a prop-2-enyl anion and a cyclopropyl anion the state correlation diagrams predict a conrotatory change for the thermal reaction, and a disrotatory change for a photochemical reaction involving the 1st excited-state.

Q5.11 Construct a state correlation diagram for the formation of cyclohexene by the suprafacial-suprafacial cyclo-addition of buta-1,3-diene and ethylene (see A3.3, page 41).

A5.11

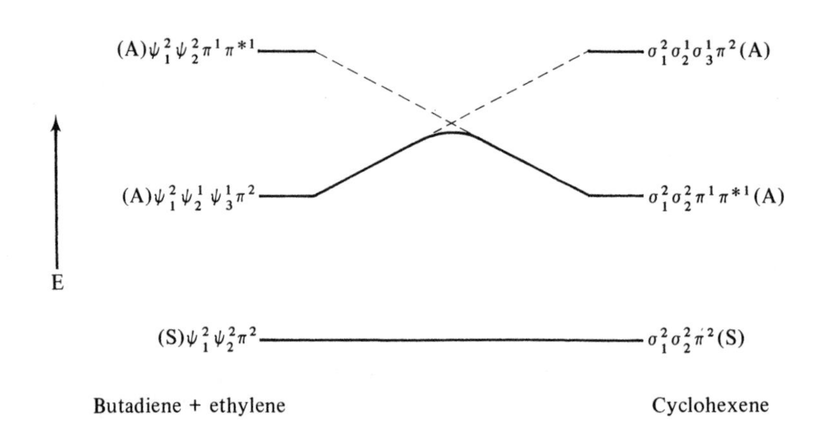

The correlations predicted by the orbital correlation diagram are shown by broken lines. Several excited-states of the interacting reactants and the product have been omitted from the diagram. The upper states shown do not correlate with each other, but with other antisymmetric states not shown.

S5.4 The correct ordering of electronic states according to their energy in the construction of a state correlation diagram is extremely important. Unfortunately, the correct order is often far from obvious for excited-states, and refined molecular orbital calculations may be necessary before a reliable decision can be made.

Further Reading

Woodward, R. B. and Hoffmann, R., The conservation of orbital symmetry, *Angew. Chem. Int. Ed.,* 1969, **8**, 781.

[Analysis of electrocyclic reactions, cyclo-addition reactions, sigmatropic migration reactions, group transfer and elimination reactions, cheletropic reactions, using frontier orbital analysis, orbital and state correlation diagrams, and the generalised selection rules for *pericyclic reactions* (a general name for all of these reactions; see p. 849), with a multitude of examples].

Dewar, M. J. S., Aromaticity and pericyclic reactions, *Angew. Chem. Int. Ed.,* 1971, **10**, 761.

Perrin, C. L., The Woodward-Hoffmann rules—an elementary approach, *Chemistry in Britain,* 1972, **8**, 163.

Zimmerman, H. E., The Möbius–Hückel concept in organic chemistry—application to organic molecules and reactions, *Acc. Chem. Res.,* 1971, **4**, 272.

[These three articles show that the transition state for an allowed thermal reaction is *aromatic,* while the transition state for an allowed photochemical reaction is *anti-aromatic*].

Salem, L., Orbital interactions and reaction paths, *Chemistry in Britain,* 1969, **5**, 449.

[A perturbational approach to the study of favourable reaction paths].

Index